ROS 机器人编程与 SLAM 算法解析指南

陶满礼 主编

人 民 邮 电 出 版 社
北 京

图书在版编目（CIP）数据

ROS机器人编程与SLAM算法解析指南 / 陶满礼主编. -- 北京 : 人民邮电出版社, 2020.2（2021.1重印）
ISBN 978-7-115-52631-1

Ⅰ. ①R… Ⅱ. ①陶… Ⅲ. ①机器人－操作系统－程序设计－指南 Ⅳ. ①TP242-62

中国版本图书馆CIP数据核字(2020)第004362号

内 容 提 要

目前 ROS（Robot Operating System）正逐步成为机器人开发领域的主要工具平台，同时 SLAM 技术也日益成为机器人应用领域的研究热点。本书主要讲解 ROS 编程与 SLAM 算法，并介绍 ROS 与机器人仿真软件 V-rep 的结合应用。书中各章节所涉及的代码均有对应的源代码，可供读者下载，便于调试与应用。

本书可作为机器人开发从业人员或相关领域研究人员的参考用书，也适合没有机器人开发基础的人员自学使用，还可作为相关培训学校的教材。

◆ 主　　编　陶满礼
　责任编辑　张　爽
　责任印制　王　郁　焦志炜

◆ 人民邮电出版社出版发行　　北京市丰台区成寿寺路 11 号
　邮编　100164　　电子邮件　315@ptpress.com.cn
　网址　http://www.ptpress.com.cn
　涿州市京南印刷厂印刷

◆ 开本：720×960　1/16
　印张：13.5
　字数：233 千字　　　　2020 年 2 月第 1 版
　印数：3 701 – 4 300 册　　2021 年 1 月河北第 5 次印刷

定价：69.00 元

读者服务热线：(010)81055410　印装质量热线：(010)81055316
反盗版热线：(010)81055315
广告经营许可证：京东市监广登字 20170147 号

前　言

本书以作者在机器人开发过程中遇到的问题及相关解决方案为基础，深入浅出地介绍了 ROS 的基础知识及其使用方法，涉及移动机器人自主建图、导航、SLAM 算法源码解读、ROS Navigation 源码解读，以及相关算法原理与代码实现。本书可帮助机器人领域的开发人员了解并使用 ROS。

本书的创新之处在于提供了结合 ROS 进行机器人开发的 V-rep 环境。V-rep 软件是一款优秀的机器人开发仿真软件。目前国内有关该软件的中文资料较少，与 ROS 结合进行开发的资料更是稀缺。本书结合 ROS 与 V-rep 这两款优秀的机器人开发软件，为读者提供了建图和导航实现的项目实例。

本书分为 7 章。

第 1 章主要介绍了 Ubuntu 系统、ROS 的安装及常用命令操作。

第 2 章主要介绍了 ROS 开发 IDE——RoboWare Studio 的安装使用，以及 ROS 开发的基础知识。

第 3 章主要介绍了 ROS 开发过程的常用调试工具。

第 4 章主要介绍了 TF 知识及使用方法。

第 5 章主要介绍了 SLAM 功能包源码解读及自定义功能包再现建图过程。

第 6 章主要介绍了 Navigation 功能包解读及相关算法原理与代码实现。

第 7 章主要介绍了 ROS 与 V-rep 联合开发实现机器人建图与导航。

本书第 1 章由夏丽娟执笔，第 2、5、6、7 章由陶满礼执笔，第 3 章由张志飞执笔，第 4 章由夏文杰执笔。全书代码由陶满礼编写并校验。

本书内容是基于作者在新松公司服务机器人开发过程中遇到的问题及相关解决方案编写而成的。新松智能交通部门工程师张志飞为作者提供了许多重要的参考意见，并参与了相关章节的编写工作，在此表示感谢！

本书逻辑清晰、合理，内容简洁、实用、易懂。对于广大从事机器人开发的工作者及开展机器人研究的高校研究人员来说，是一本不可多得的自学及参考用书。

由于编者水平有限，书中难免存在错误和欠缺之处，敬请读者批评指正。

编者

2019 年 9 月

资源与支持

本书由异步社区出品，社区（https://www.epubit.com/）为您提供相关资源和后续服务。

配套资源

本书提供配套源代码，请在异步社区本书页面中点击 配套资源 ，跳转到下载界面，按提示进行操作即可。注意：为保证购书读者的权益，该操作会给出相关提示，要求输入提取码进行验证。

提交勘误

作者和编辑尽最大努力来确保书中内容的准确性，但难免会存在疏漏。欢迎您将发现的问题反馈给我们，帮助我们提升图书的质量。

当您发现错误时，请登录异步社区，按书名搜索，进入本书页面，点击“提交勘误”，输入勘误信息，点击“提交”按钮即可。本书的作者和编辑会对您提交的勘误进行审核，确认并接受后，您将获赠异步社区的100积分。积分可用于在异步社区兑换优惠券、样书或奖品。

扫码关注本书

扫描下方二维码，您将会在异步社区微信服务号中看到本书信息及相关的服务提示。

与我们联系

我们的联系邮箱是 contact@epubit.com.cn。

如果您对本书有任何疑问或建议，请您发邮件给我们，并请在邮件标题中注明本书书名，以便我们更高效地做出反馈。

如果您有兴趣出版图书、录制教学视频，或者参与图书翻译、技术审校等工作，可以发邮件给我们；有意出版图书的作者也可以到异步社区在线提交投稿（直接访问 www.epubit.com/selfpublish/submission 即可）。

如果您是学校、培训机构或企业，想批量购买本书或异步社区出版的其他图书，也可以发邮件给我们。

如果您在网上发现有针对异步社区出品图书的各种形式的盗版行为，包括对图书全部或部分内容的非授权传播，请您将怀疑有侵权行为的链接发邮件给我们。您的这一举动是对作者权益的保护，也是我们持续为您提供有价值的内容的动力之源。

关于异步社区和异步图书

“异步社区”是人民邮电出版社旗下 IT 专业图书社区，致力于出版精品 IT 技术图书和相关学习产品，为作译者提供优质出版服务。异步社区创办于 2015 年 8 月，提供大量精品 IT 技术图书和电子书，以及高品质技术文章和视频课程。更多详情请访问异步社区官网 https://www.epubit.com。

“异步图书”是由异步社区编辑团队策划出版的精品 IT 专业图书的品牌，依托于人民邮电出版社近 30 年的计算机图书出版积累和专业编辑团队，相关图书在封面上印有异步图书的 LOGO。异步图书的出版领域包括软件开发、大数据、AI、测试、前端、网络技术等。

异步社区

微信服务号

目 录

第1章 ROS简介

1.1 ROS概述

ROS，全称 Robot Operating System，是一个开源的机器人操作系统，能为异质计算机集群提供类似操作系统的功能（注意：ROS 不是真正意义上的操作系统，它通常运行在 Ubuntu 系统上，且有固定的版本对应）。ROS 提供了操作系统应有的服务，包括硬件抽象、底层设备控制、常用函数的实现、进程间消息传递，以及包管理。ROS 也提供用于获取、编译、编写和跨计算机运行代码所需的工具和库函数。

ROS 系统起源于 2007 年斯坦福大学人工智能实验室与机器人技术公司 Willow Garage 合作的个人机器人项目 PR2（Personal Robots Program）。2009 年初推出测试版的 ROS0.4，该版本已初步具备现有的 ROS 系统框架。2010 年正式推出 ROS1.0 版本，并开发出一系列机器人操作的基础软件包。之后不断进行版本迭代和功能完善，目前 ROS 的最新版本为 Lunar。除此之外，支持包括 Linux、Windows、macOS 等操作系统的 ROS2 也已推出（本书只介绍 ROS1 的使用，对 ROS2 不作讨论）。ROS 各版本及其发布时间如表 1-1 所示。

表 1-1　ROS 版本及发布时间

ROS 版本	发布时间
Lunar Loggerhead	2017.5
Kinetic Kame	2016.5
Jade Turtle	2015.5
Indigo Igloo	2014.7
Hydro Medusa	2013.9
Groovy Galapagos	2012.12
Fuerte Turtle	2012.4
Electric Emys	2011.8
Diamondback	2011.3
C Turtle	2010.8
Box TurtleBox Turtle	2010.3

1.2 Ubuntu系统

Ubuntu（乌班图）是一个基于 Debian 的以桌面应用为主的 Linux 操作系统，支持 x86、amd64（即 x64）架构，由全球化的专业开发团队（Canonical Ltd）打造。机器人操作系统 ROS 就是基于 Ubuntu 运行的，因此在学习 ROS 之前，我们需要先花一些时间来了解如何安装以及配置 Ubuntu 系统，为之后安装与配置 ROS 系统做准备。不同的 ROS 版本对应不同版本的 Ubuntu 系统，其匹配关系如表 1-2 所示。

表 1-2　ROS 版本及对应版本的 Ubuntu 系统

ROS 发布日期	ROS 版本	Ubuntu 系统版本
2016.3	ROS Kinetic Kame	Ubuntu 16.04 (Xenial) / Ubuntu 15.10 (Wily)
2015.3	ROS Jade Turtle	Ubuntu 15.04 (Wily) / Ubuntu LTS 14.04 (Trusty)
2014.7	ROS Indigo Igloo	Ubuntu 14.04 (Trusty)
2013.9	ROS Hydro Medusa	Ubuntu 12.04 LTS (Precise)
2012.12	ROSGroovyGalapagos	Ubuntu 12.04

1.2.1 Ubuntu系统的安装

准备工作如下。

- Ubuntu 的镜像文件。
- U 盘，用于制作启动盘。
- UltraISO 软件，用于刻录启动 U 盘。

进入 Ubuntu 官网 https://www.ubuntu.com/download/alternative-downloads 下载安装包，界面显示可供下载的 Ubuntu 镜像选项如下。

- Download the network installer for 18.10
- Download the network installer for 18.04 LTS
- Download the network installer for 16.04 LTS
- Download the network installer for 14.04 LTS

选择下载对应版本（本书以 14.04 为例），后续对应 ROS 的 Indigo 版本使用。为电脑分出 30GB ～ 70GB 的存储空间，从现有的硬盘中直接压缩。具体操作是，右击“计算机→管理→磁盘管理”，可以很清楚地看到各个磁盘的分区情况，右击选中待压缩的磁盘，单击压缩卷，压缩出 30GB ～ 70GB 的内存用于安装 Ubuntu 系统。

完成分区之后开始刻录 U 盘启动盘，安装下载好的 UltraISO 软件并打开，如图 1-1 所示，在菜单里找到“启动”选项，单击“写入硬盘映像”。

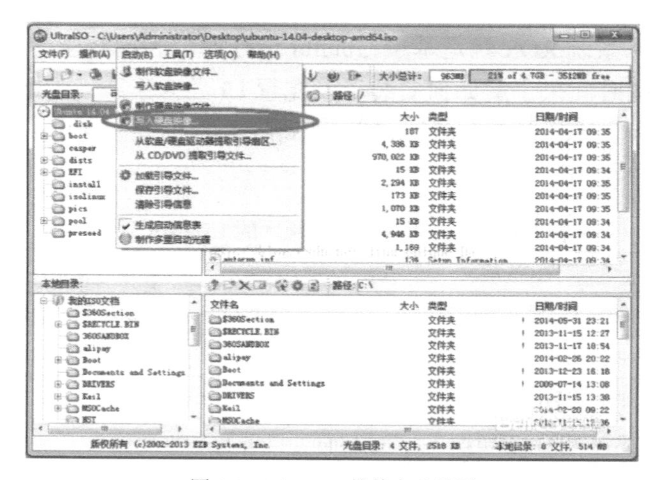

图 1-1　UltraISO 软件启动界面

在弹出的窗口中单击“便捷启动”选项，在下拉菜单中选择“写入新的硬盘主引导记录（MBR）”，继续选择“USB-HDD+”，如图 1-2 所示。

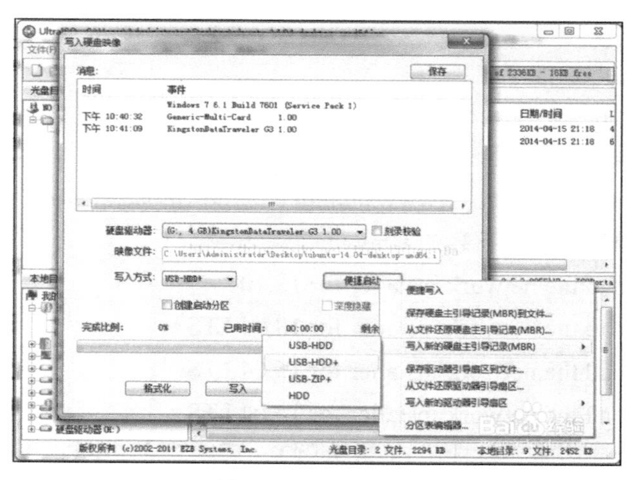

图 1-2　写入磁盘映像

写入完成后，拔出 U 盘并关闭电脑，然后重新插入 U 盘，启动电脑，通过快捷键进入 BIOS，选择 U 盘启动。各电脑进入 BIOS 的快捷键可参考表 1-3。

表 1-3 BIOS 快捷键参考

笔记本	启动按键	台式机	启动按键
联想笔记本	F12	联想台式机	F12
宏基笔记本	F12	惠普台式机	F12
华硕笔记本	Esc	宏基台式机	F12
惠普笔记本	F9	戴尔台式机	Esc
联想 ThinkPad	F12	神舟台式机	F12
戴尔笔记本	F12	华硕台式机	F8
神舟笔记本	F12	方正台式机	F12
东芝笔记本	F12	清华同方台式机	F12
三星笔记本	F12	明基台式机	F8
IBM 笔记本	F12		

进入 BIOS 之后，在安装界面选择“中文（简体）”，如图 1-3 所示，单击“安装 Ubuntu”。

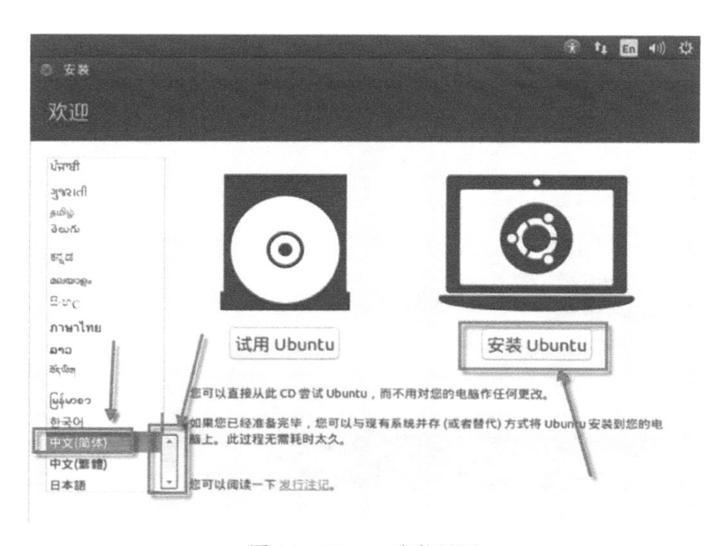

图 1-3 Ubuntu 安装界面

在准备安装 Ubuntu 界面中单击“继续”，如图 1-4 所示。

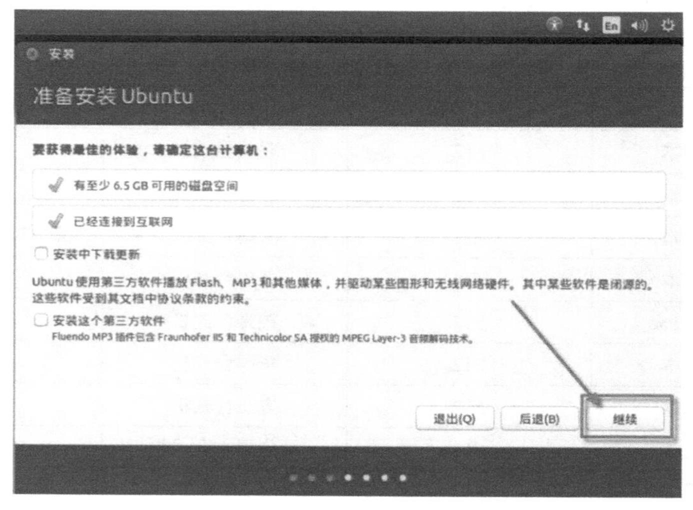

图 1-4　准备安装界面

在新弹出的安装类型界面中选择“其他选项”，单击“继续”，如图 1-5 所示。

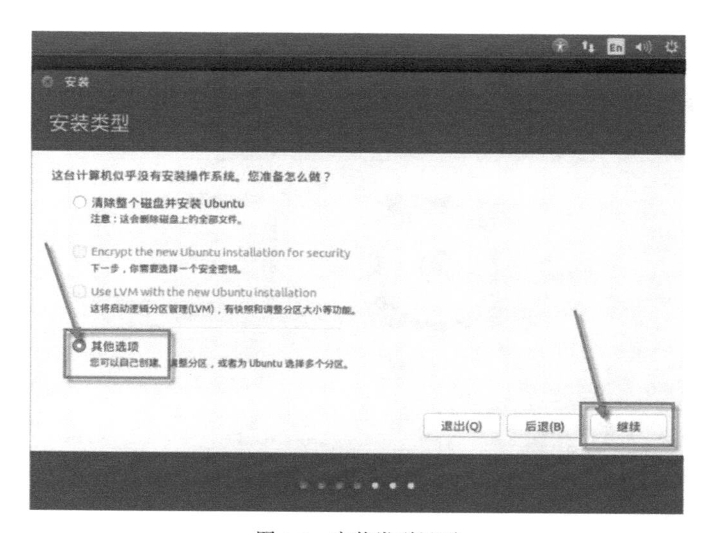

图 1-5　安装类型界面

在弹出的新建分区界面中，单击“新建分区表 ...”，如图 1-6 所示。

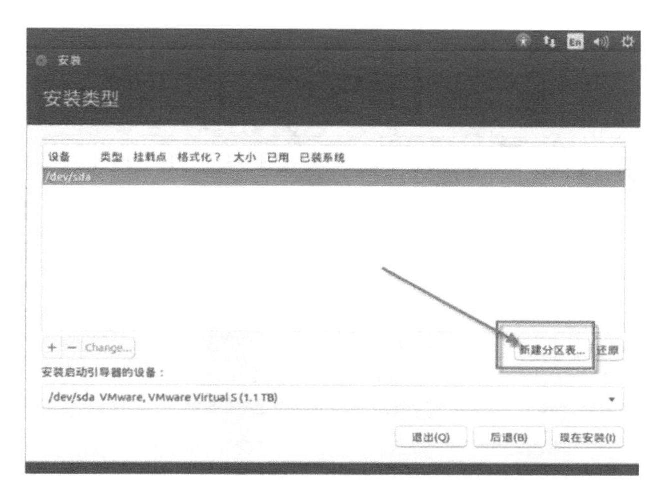

图 1-6　新建分区界面

此时，弹出窗口提示“要在此设备上创建新的空分区表吗？”，选择“继续”，如图 1-7 所示。

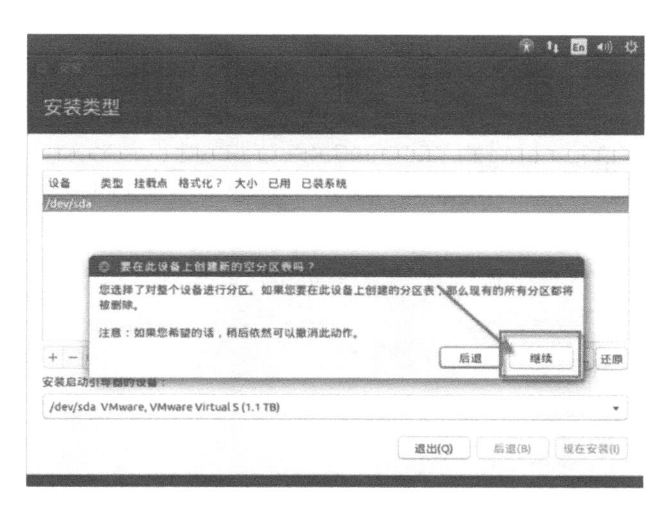

图 1-7　是否新建分区表界面

接下来，建立 /boot 分区。如图 1-8 所示，单击“空闲”磁盘分区，单击“+”添加新分区，将大小设置为 500MB，类型为“主分区”，挂载点为“/boot”，最后单击“确定”。

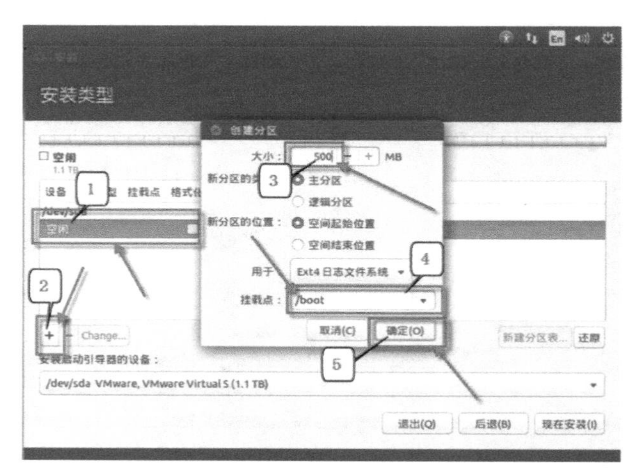

图 1-8　新建 boot 分区

接下来，新建交换空间分区。如图 1-9 所示，单击“空闲”磁盘分区，单击“+”添加新分区，将大小设置为 2048MB，类型为“主分区”，用于“交换空间”，最后单击“确定”。

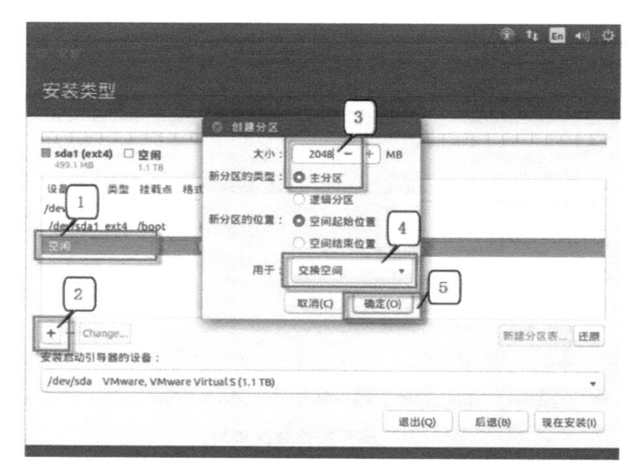

图 1-9　新建交换空间分区

然后，新建根分区。如图 1-10 所示，单击“空闲”磁盘分区，单击“+”添加新分区，将大小设置为 50GB，类型为“主分区”，挂载点为“/”根分区，最后单击“确定”。

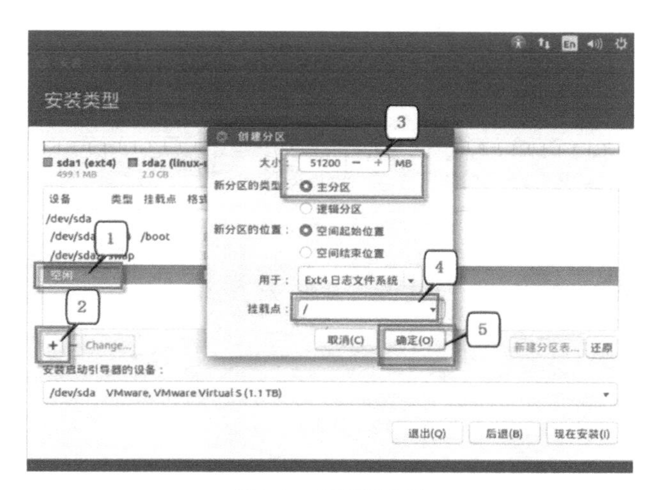

图 1-10 新建根分区

接着，新建 home 分区。单击“空闲”磁盘分区，单击“+”添加新分区，将大小设置为剩余全部空间，类型为“逻辑分区”，挂载点为“/home”，单击“确定”，最后单击“现在安装”，如图 1-11 所示。

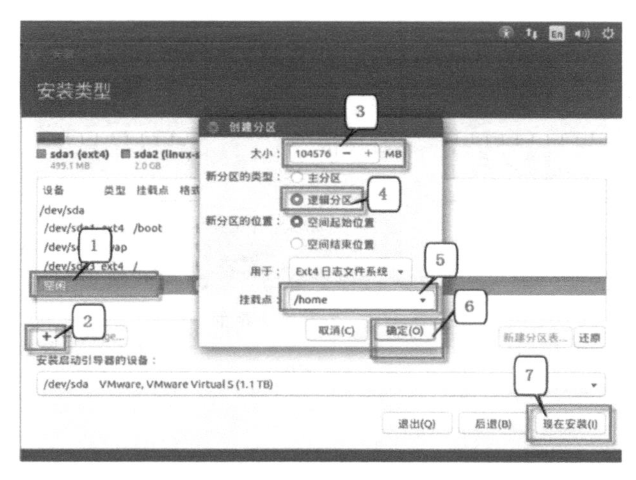

图 1-11 新建 home 分区

接下来，在页面中设置用户名、计算机名及密码，单击“继续”，安装完成后单击“现在重启”，如图 1-12 所示。

图 1-12　是否重启界面

重启后，可看到图 1-13 所示的安装成功界面。

图 1-13　安装成功界面

至此，Ubuntu 系统在电脑上的安装工作完成。

1.2.2　树莓派安装Ubuntu

1. 准备工作

（1）树莓派：HDMI 转 VGA 转接线（连接树莓派与显示器）、电源（树莓派供电）、microSD 卡（安装系统）。.img 镜像的官方下载地址为 http://cdimage.

ubuntu.com/ubuntu/releases/。

（2）Win32diskmanager（镜像读写工具）的官方下载地址为 https://sourceforge.net/projects/win32diskimager/。

2. 安装步骤

（1）将 microSD 卡插入读卡器。

（2）打开 Win32DiskImager，选择下载的 .img 镜像及 microSD 卡的盘符，然后写入。

（3）将写好的系统内存卡插入树莓派，然后将键盘、鼠标、显示器、树莓派插上电源即可启动。

（4）设置系统语言、用户名和密码等。

1.3 ROS的安装

配置 Ubuntu 软件仓库，打开软件和更新对话框，具体可以在 Ubuntu 最左上角的“搜索”选项中搜索。打开后按照图 1-14 所示进行配置（确保勾选了“restricted”“universe”和“multiverse”）。

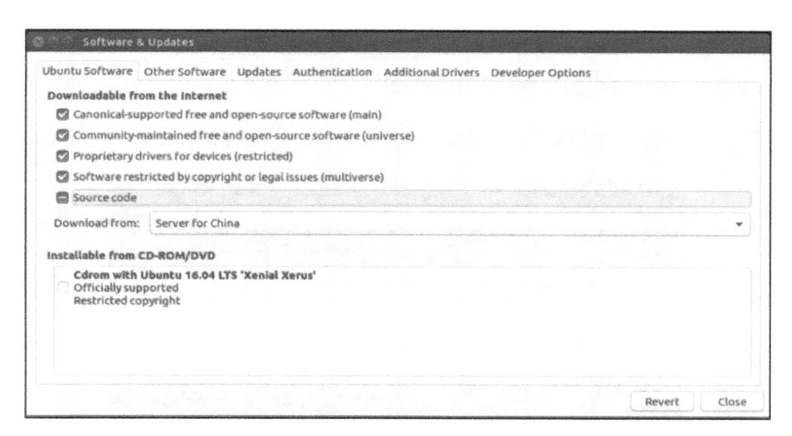

图 1-14　配置 Ubuntu 软件仓库

添加 sources.lists，配置电脑使其能够安装来自 packages.ros.org 的软件。ROS Indigo 仅支持 Saucy (13.10) 和 Trusty (14.04)：

```
sudo sh -c 'echo "deb http://packages.ros.org/ros/ubuntu $(lsb_release -sc) main" > /etc/apt/sources.list.d/ros-latest.list'
```

添加 keys：

```
sudo apt-key adv --keyserver hkp://pool.sks-keyservers.net --recv-key 421C365B
D9FF1F717815A3895523BAEEB01FA116
```

更新 Debian 软件包索引：

```
sudo apt-get update
```

安装桌面完整版 ROS：

```
sudo apt-get install ros-indigo-desktop-full
```

在开始使用 ROS 之前，还需要初始化 rosdep。rosdep 可以在编译某些源码时为其安装一些系统依赖，也是某些 ROS 核心功能组件必须要用到的工具。

```
sudo rosdep init
rosdep update
```

下面开始配置环境。

打开终端，输入以下内容：

```
$ echo "source /opt/ros/kinetic/setup.bash" >> ~/.bashrc
$ source ~/.bashrc
```

安装 rosinstall。rosintall 是 ROS 中经常使用的命令行，可以便于我们在一行命令中同时下载多个 ROS packages，其安装命令如下：

```
sudo apt-get install python-rosinstall
```

测试时，在命令行输入“roscore”启动 ROS 节点管理器，如图 1-15 所示。

```
... logging to /home/ubuntu/.ros/log/972a7194-195c-11e9-83b1-0c84dc946419/roslau
nch-ubuntu-Lenovo-IdeaPad-Y480-28399.log
Checking log directory for disk usage. This may take awhile.
Press Ctrl-C to interrupt
Done checking log file disk usage. Usage is <1GB.

started roslaunch server http://192.168.43.187:46230/
ros_comm version 1.11.21


SUMMARY
========

PARAMETERS
 * /rosdistro: indigo
 * /rosversion: 1.11.21

NODES

auto-starting new master
process[master]: started with pid [28451]
ROS_MASTER_URI=http://192.168.43.187:11311/

setting /run_id to 972a7194-195c-11e9-83b1-0c84dc946419
process[rosout-1]: started with pid [28464]
started core service [/rosout]
```

图 1-15　ROS 节点管理器启动界面

如果看到图 1-15 所示的结果，说明已成功安装 ROS。

1.4 常用的操作命令

1.4.1 Ubuntu系统的常用命令

由于 ROS 是基于 Ubuntu 系统运行的，因此在学习 ROS 之前，我们需要熟悉一些基本的 Linux 操作指令。常用的操作命令分为以下 3 类。

（1）查找指令

- ls：列出当前目录下的所有文件（不显示隐藏文件）。
- ls –a：列出当前目录下的所有文件（显示隐藏文件）。
- pwd：显示当前路径。
- cat /etc/issue：查看 Ubuntu 版本。
- sudofdisk –l：查看磁盘信息。
- free –m：查看当前的内存使用情况。
- ps –A：查看当前有哪些进程。

（2）文件 / 文件夹操作命令

- cd ~/file_path：转到（切换）路径。
- mkdir dirname：新建目录。
- rmdir dirname：删除空目录。
- rm filename：删除文件。
- rm –rf dirname：删除非空目录及其包含的所有文件。
- mv file1file2：将文件 1 重命名为文件 2。
- mv file1 dir1：将文件 1 移动到目录 1。

（3）安装软件及创建用户命令

- sudo apt-get install xxx：安装程序。
- sudo apt-get remove –purge xxx：彻底卸载程序并清空配置。
- sudo apt-get update：更新本地软件源文件。
- sudo user add username：创建一个新的用户 username。
- sudo passwd username：设置用户 username 的密码。
- sudo group add groupname：创建一个新的 groupname 组。

- sudo usermod –g groupname username：把用户 username 加入 groupname 组中。

1.4.2 常用的ROS操作命令

ROS 中包含丰富的调试命令。本书只提供一些常用的基本操作命令，这些命令会在后续章节中频繁地被使用，希望读者熟记。

（1）工作空间及功能包相关命令

- rosc：切换（cd）工作目录到某个程序包（或其子目录）roscd [package[/subdir]]。
- rosls：按程序包的名称执行 ls 命令 rosls [package[/subdir]]。
- catkin_create_pkg：创建功能包。
- catkin_make：编译 ROS 工作空间。

（2）节点启动及主题调试命令

- roscore：启动管理器。
- rosrun：运行 ROS 包中的一个可执行文件 rosrun package_name executable_name。
- ruslaunch：启动 roscore、本地节点和远程节点（通过 ssh），设置参数服务器的参数。
- roslaunch package_name file_name.launch：启动包中的一个文件。
- rospack：获取程序包的有关信息。
- rospack find [package]：返回程序包的路径。
- rospack list：获取所有的程序包。
- rosdep：一个能够下载并安装 ROS packages 所需要的系统依赖项的工具 rosdep install [package]。
- roswtf：可以检查 ROS 系统并尝试发现问题，如 roswtf or roswtf [file]。
- rostopic –h：查看 rostopic 的所有操作。
- rostopic list：查看所有 Topic 列表。
- rosrun rqt_plot rqt_plot：图形化显示 Topic。
- rostopic echo [topic]：查看某个 Topic 信息。

第 2 章 ROS 基础

2.1 开发工具IDE

RoboWare Studio 是一个 ROS 集成开发环境，使 ROS 开发更加直观、简单，并且易于操作，可用于 ROS 工作区及包的管理，代码编辑、构建及调试。RoboWare Studio 专为 ROS（indigo/jade/kinetic）设计，以图形化的方式完成 ROS 工作区及包的创建、源码添加、message/service/action 文件创建、显示包及节点列表，并且可以实现 CMakelists.txt 文件和 package.xml 文件的自动更新。下面介绍该开发环境的安装与使用。

2.1.1 RoboWare Studio的安装

1. 准备工作

（1）操作系统为 Ubuntu。

（2）已完成 ROS 的安装配置。

（3）可以使用 catkin_make 构建 ROS 包（若无法构建，可以运行下列命令来安装构建工具）：

```
sudo apt-get install build-essential
```

（4）安装 python-pip：

```
sudo apt-get install python-pip
```

2. 安装

进入 RoboWare 官网，界面如图 2-1 所示。

选择下载对应的 RoboWare Studio 版本，双击下载下来的 .deb 文件即可完成安装。安装过程中若弹出用户协议，则直接按 <Esc> 键；若弹出“您是否接受上述协议？”窗口，则单击“确定”，按回车键，自动开始安装，如图 2-2 所示。

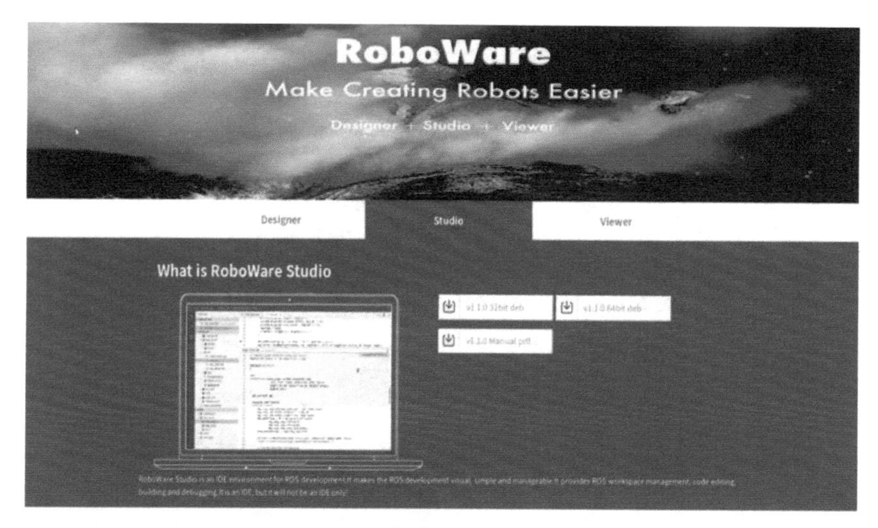

图 2-1　RoboWare 官网界面

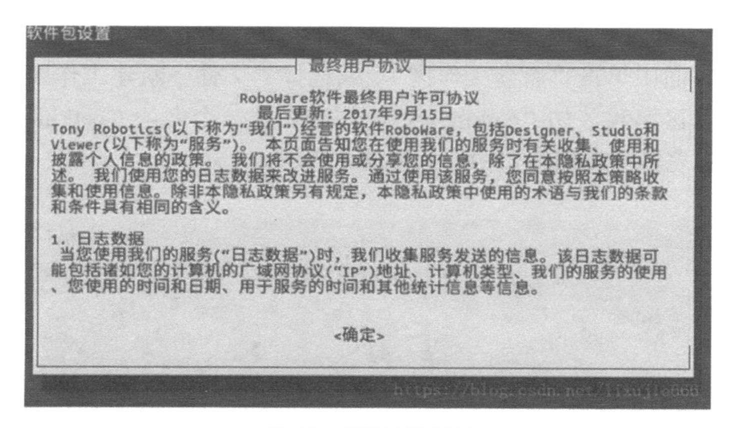

图 2-2　用户协议界面

安装成功后，可看到图 2-3 所示的工作界面。

2.1.2　卸载

打开任意一个终端，执行以下指令来卸载 RoboWare Studio：

```
sudo apt-get remove roboware-studio
```

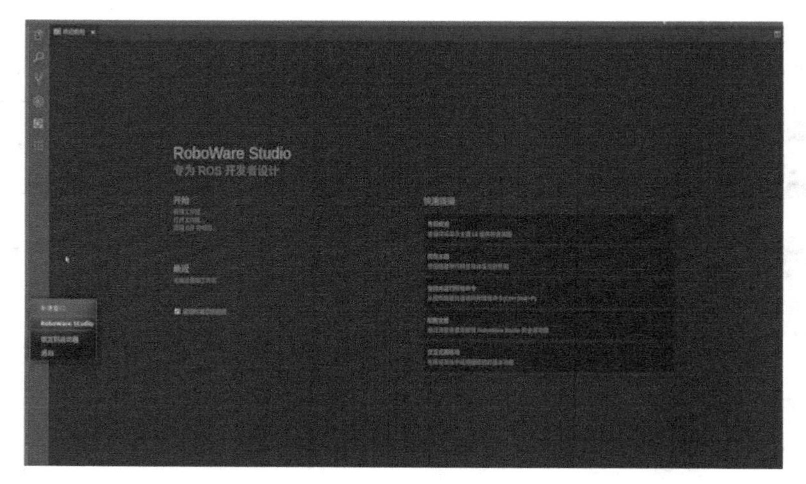

图 2-3　工作界面

2.1.3　使用

1. 创建工作区

在欢迎界面中单击“新建工作区”（或选择“文件→新建工作区”），选择路径并填写工作区名称，如“catkin_ws”，则会创建一个名为“catkin_ws”的工作区，并显示在资源管理器窗口中，如图 2-4 所示。

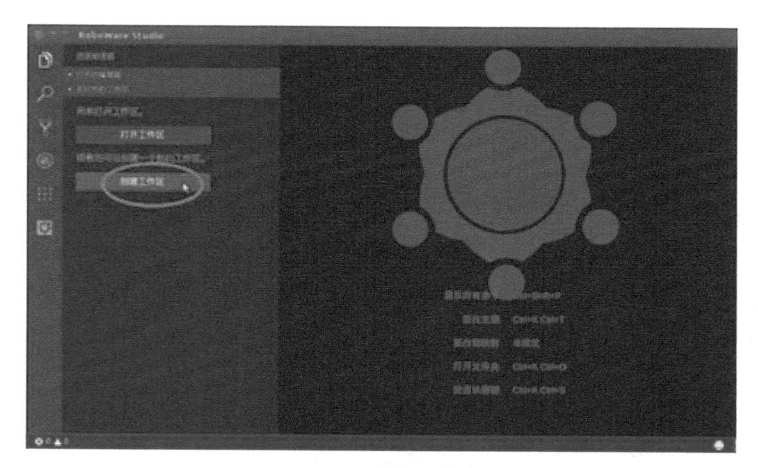

图 2-4　创建工作区

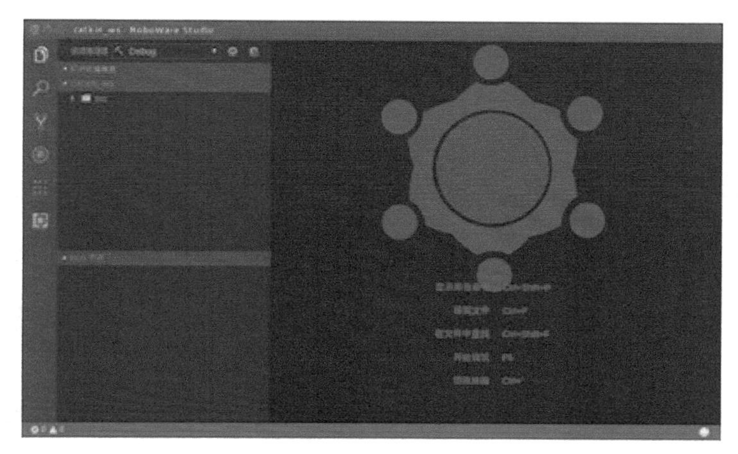

图 2-4　创建工作区（续）

2. 创建 ROS 包

右击 ROS 工作区下的“src”，选择“新建 ROS 包”，输入包名称及其依赖包的名称。按回车键后，会创建对应的 ROS 包，如图 2-5 所示。

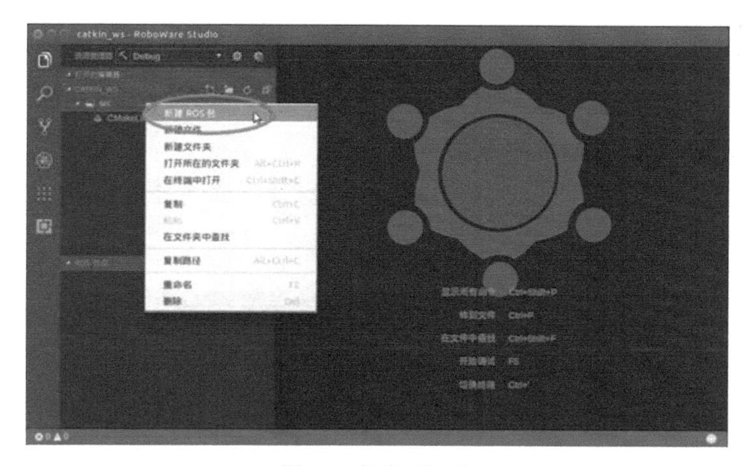

图 2-5　创建 ROS 包

右击 ROS 包目录下的“src”，选择“新建 CPP 源文件”，输入文件名后，按回车键，会弹出列表。在列表中选择类型“加入新的可执行文件中”，则会创建一个与 CPP 文件同名的执行文件（ROS 节点），此时 CMakeLists.txt 文件会自动更新。

右击包名文件夹，依次选择“新建 Msg 文件夹”“新建 Srv 文件夹”“新建 Action 文件夹”，可分别创建 Message、Service、Action 文件夹，右击相应文件夹即可添加 Message、Service、Action 文件。此时 CMakeLists.txt 文件会自动更新，如图 2-6 所示。

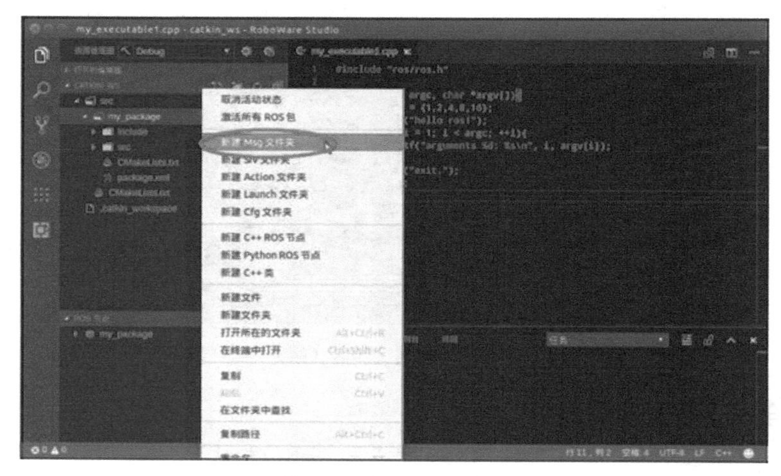

图 2-6　新建 Msg 文件夹

3. 编译文件

单击左上角的小锤子图标，进行编译文件，或者通过在菜单栏中单击“ROS → build”进行编译。

4. 添加并启动 Launch 文件

首先，右击包名文件夹（如“my_package”），选择“新建 Launch 文件夹”来创建 Launch 文件夹，如图 2-7 所示。

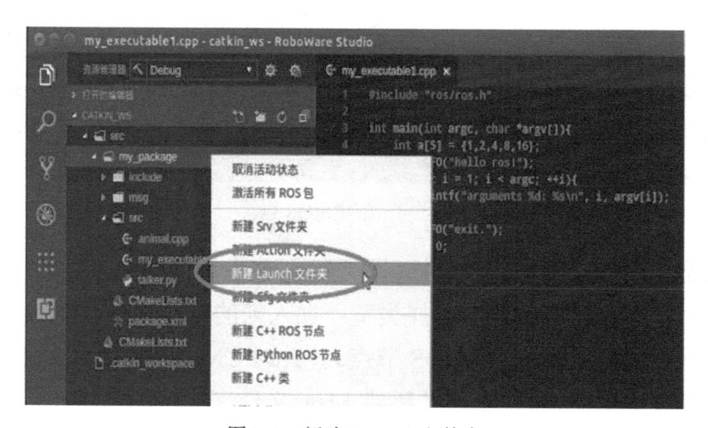

图 2-7　新建 Launch 文件夹

然后，右击 Launch 文件夹，选择“新建 LAUNCH 文件”，输入文件名添加 Launch 文件，如图 2-8 所示。

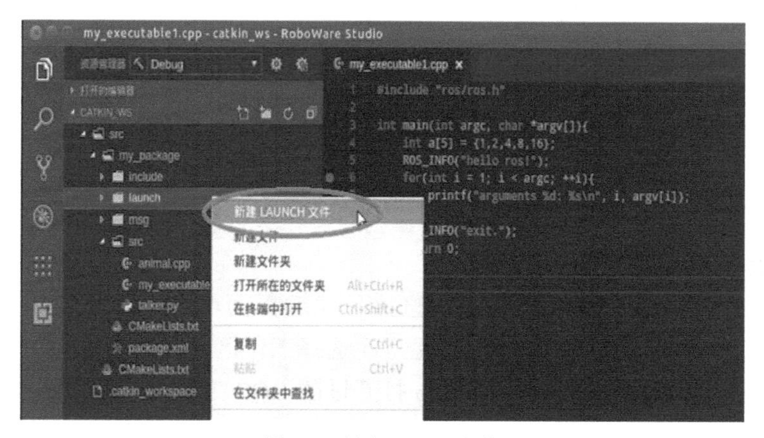

图 2-8　新建 Launch 文件

编辑完成后，右击 Launch 文件，选择“运行 Launch 文件”，RoboWare Studio 就会自动打开集成终端并运行 Launch 文件，如图 2-9 所示。

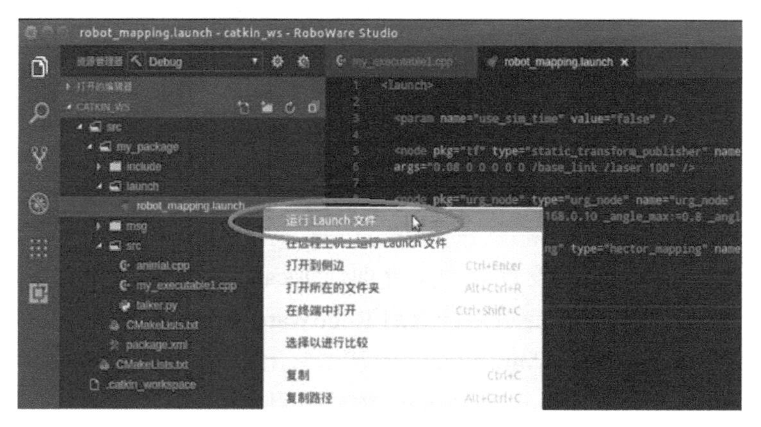

图 2-9　运行 Launch 文件

在菜单栏中，依次选择“ROS- 运行 roscore”“ROS- 运行 RViz”“ROS- 运行 rqt”“ROS- 运行 rqt-reconfigure”“ROS- 运行 rqt-graph”，可分别启动 roscore、RViz、rqt、rqt-reconfigure、rqt-graph。图 2-10 所示为启动 rqt-reconfigure 的界面。

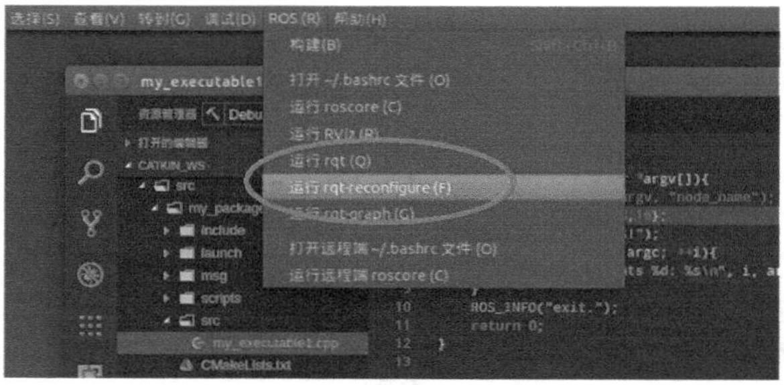

图2-10　启动rqt-reconfigure

5. 导入已有的ROS工作区

（1）对于普通的ROS工作区，直接在欢迎界面单击“打开工作区”按钮，或在菜单栏中选择“文件→打开工作区”，选择工作区路径打开。

（2）对于旧版本RoboWare Studio打开过的ROS工作区，需要将工作区根目录下的“.vscode”文件夹删除，再打开工作区。

6. 修改界面语言

在菜单栏中，选择“文件→首选项→语言设置”，打开配置文件，如图2-11所示。

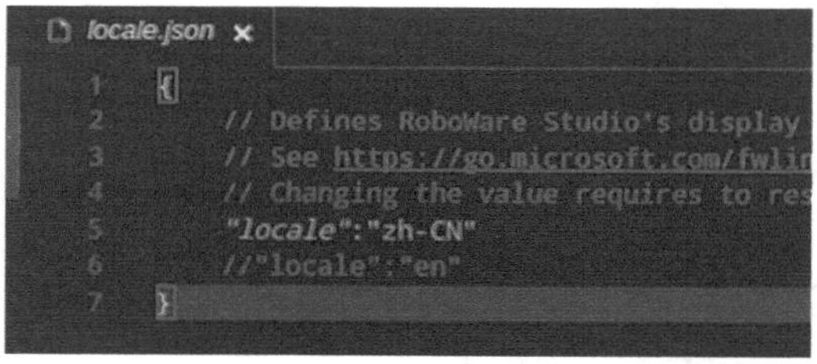

图2-11　语言设置

"locale":"zh-CN"表示设置为中文界面，"locale":"en"表示设置为英文界面，可用“//”进行注释。修改完成后，重启RoboWare Studio即可生效。

这里我们只介绍RoboWare Studio的基本常用方法。读者若对软件的具体使用感兴趣，可自行查阅RoboWare Studio软件使用手册。

2.2 节点

在ROS中，最小的进程单位称为节点（Node）。每个节点都是一个可执行文件，负责实现具体的功能，如激光数据获取处理、坐标变换、图像信息获取处理里程及发布等。构成节点的可执行文件通常由 C++ 和 Python 等语言编写。节点通过在具体主题上发布对应消息与其他节点进行通信。此外，ROS 也通过节点管理器 roscore 来管理运行中的各节点。下面以接收端和发布端为例，通过编写相应节点来阐述 ROS 节点间的通信原理。

2.2.1 发布端

打开编辑器，创建发布端 PubForBeginner.cpp，内容如下：

```
#include "ros/ros.h"
#include "std_msgs/String.h"

int main(int argc,char** argv)
{
    //initialize and name node
   ros::init(argc,argv,"publisher");    //create nodehandle
    ros::NodeHandle nh;
    //create publisher
    ros::Publisher simplepub = nh.advertise<std_msgs::String>("string_topic",100);
    // publish frequency
    ros::Rate rate(10);
    //message for publish
    std_msgs::String pubinfo;
    pubinfo.data="Hello, I'm Publisher!";
    while(ros::ok())
    {
        simplepub.publish(pubinfo);
        rate.sleep();
    }
    return 0;
}
```

说明如下。

- ros::init(argc,argv,"publisher")：创建 ros 节点，并命名为 publisher。
- ros::NodeHandle nh：创建 ros node handle 句柄。
- ros::Publisher simplepub = nh.advertise<std_msgs::String>("string_topic",100)：创建发布者，并在 string_topic 主题上发布 std_msgs::String 类型的消息。
- pubinfo.data="Hello, I'm Publisher!"：消息内容。

- simplepub.publish(pubinfo)：结合 while 和 ros::Rate 按照固定频率循环发布消息。

2.2.2 接收端

打开编辑器，创建接收端 SubForBeginner.cpp，内容如下：

```
#include "ros/ros.h"
#include "std_msgs/String.h"

using namespace std;

void subCallback(const std_msgs::String& submsg)
{
    string subinfo;
    subinfo = submsg.data;
    ROS_INFO("The message subscribed is: %s",subinfo.c_str());
}
int main(int argc,char** argv)
{
    //initial and name node
    ros::init(argc,argv,"subscriber");
    //create nodehandle
    ros::NodeHandle nh;
    //create subscriber
    ros::Subscriber sub = nh.subscribe("string_topic",1000,&subCallback);
    ros::spin();
    return 0;
}
```

说明如下。

- void subCallback(const std_msgs::String& submsg)：订阅器回调函数，对接收到的消息进行打印处理。
- ros::Subscriber sub = nh.subscribe("string_topic",1000,&subCallback)： 订阅器，订阅 string_topic 主题上的消息，并通过回调函数进行处理。
- ros::spin()：ROS 消息回调处理函数，程序运行到这里后不再往下执行。

2.2.3 CMakeLists.txt文件

在 CMakeLists.txt 文件中添加如下内容：

```
    cmake_minimum_required(VERSION 2.8.3)
    project(book_node)
    find_package(catkin REQUIRED COMPONENTS roscpp rospy std_msgs)
    catkin_package(
    )
    include_directories(
      include ${catkin_INCLUDE_DIRS}
    )
    add_executable(PubForBeginner
      src/PubForBeginner.cpp
    )
    add_dependencies(PubForBeginner ${${PROJECT_NAME}_EXPORTED_TARGETS} ${catkin_
EXPORTED_TARGETS})
    target_link_libraries(PubForBeginner
      ${catkin_LIBRARIES}
    )
    add_executable(SubForBeginner
      src/SubForBeginner.cpp
    )
    add_dependencies(SubForBeginner ${${PROJECT_NAME}_EXPORTED_TARGETS} ${catkin_
EXPORTED_TARGETS})
    target_link_libraries(SubForBeginner
      ${catkin_LIBRARIES}
    )
```

编译即可获得 PubForBeginner 和 SubForBeginner 可执行文件。

2.2.4 测试

打开两个终端分别运行：

```
rosrun book_node SubForBeginner
rosrun book_node PubForBeginner
```

可以在接收端看到如下打印信息：

```
[ INFO] [1551682316.518305766]: The message subscribed is: Hello, I'm Publisher!
[ INFO] [1551682316.618326302]: The message subscribed is: Hello, I'm Publisher!
[ INFO] [1551682316.718301812]: The message subscribed is: Hello, I'm Publisher!
```

在新终端运行 rosrun rqt_graph rqt_graph，可以查看节点间的通信关系，如图 2-12 所示。

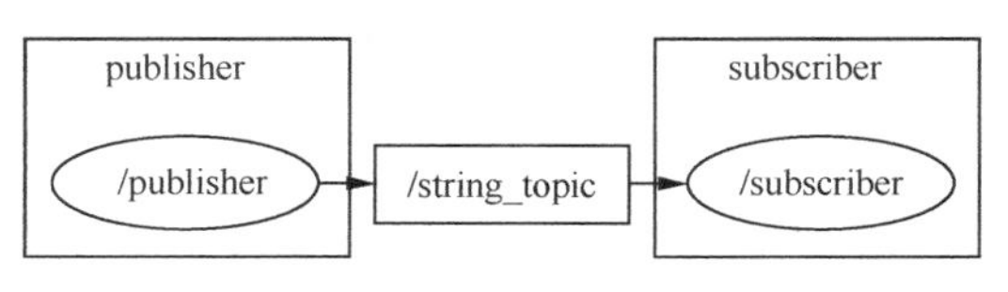

图 2-12 节点间的通信关系

图 2-12 清晰地表明，publisher 节点和 subscriber 节点通过 string_topic 主题成功实现通信。

2.3 消息

ROS 中有许多已定义的消息类型，如 int32、int64、float64 和 string 等。但是为了满足不同场景功能需求，通常需要用户自定义消息类型。下面我们将练习如何自定义消息类型，并进行发布和接收。

2.3.1 自定义消息

在 RoboWare Studio 创建通过 Add Msg Folder 创建 msg 文件夹，通过 Add Msg File 创建 Student.msg 文件，内容如下。

- string name
- float64 height
- float64 weight

该消息主要包含 3 部分。

- 字符串类型的学生姓名信息。
- float64 类型的学生身高信息。
- float64 类型的学生体重信息。

在 CMakeLists.txt 文件中添加如下内容：

```
find_package(catkin REQUIRED COMPONENTS
  message_generation roscpp rospy std_msgs)

add_message_files(FILES
  Student.msg
)
```

```
generate_messages(DEPENDENCIES
  std_msgs
)
```

编译之后，在 .../robware_ws/devel/include/book_message 文件夹下可看到自动生成的头文件：Student.h。

2.3.2 编写自定义消息发布端

新建文件 Book_MyMsgPub.cpp，内容如下：

```
#include "ros/ros.h"
#include "book_message/Student.h"
#include <cstdlib>

using namespace std;
int main(int argc,char** argv)
{
    //initial and name node
    ros::init(argc,argv,"node_MyMsgPub");
    if(argc!=4)
    {
        cout<<"Error command parameter!\n"\
        <<"Please run command eg:\n"\
        <<"rosrun book_messge Book_MyMsgPub LiLei 180 160"<<endl;
        return 1;
    }

    //create node handle
    ros::NodeHandle nh;

    //create message publisher
    ros::Publisher MyMsgPub =  nh.advertise<book_message::Student>("MyMsg",100);
    book_message::Student  sdtMsg;
    sdtMsg.name = argv[1];
    //convert string in argv to float and pass value to height, weight
    sdtMsg.height = atof(argv[2]);
    sdtMsg.weight = atof(argv[3]);

    ros::Rate rate(10);
    while(ros::ok())
    {
```

```
        MyMsgPub.publish(sdtMsg);
        rate.sleep();
    }

    return 0;
}
```

说明如下。

- ros::init(argc,argv,"node_MyMsgPub")：初始化并命名节点名称。
- if(argc!=4)：该段代码的主要功能是提醒使用者如何使用该节点。
- ros::Publisher MyMsgPub = nh.advertise<book_message::Student>("MyMsg", 100)：在 MyMsg 主题上发布用户自定义的 book_message::Student 类型消息。
- book_message::Student sdtMsg：定义将发布的消息。
- sdtMsg.name = argv[1]：命令行第二个参数作为学生的名字信息。
- sdtMsg.height = atof(argv[2])：命令行第三参数作为学生的身高信息（atof() 函数将字符串转化为浮点数）。
- sdtMsg.weight = atof(argv[3])：命令行第四个参数作为学生的体重信息。
- MyMsgPub.publish(sdtMsg)：循环体中以固定频率发布消息。

2.3.3 编写自定义消息接收端

创建接收端文件 Book_MyMsgSub.cpp，内容如下：

```
#include "ros/ros.h"
#include "book_message/Student.h"

// custom defined message callback function
void MyMsgCallback(const book_message::Student& sdtInfo)
{
    ROS_INFO("The student information is:\n name:%s---height:%f---weight:%f",
             sdtInfo.name.c_str(),
             sdtInfo.height,
             sdtInfo.weight);
}

int main(int argc, char** argv)
{
    //initial and name node
```

```
    ros::init(argc,argv,"node_MyMsgSub");
    //create node handle
    ros::NodeHandle nh;
    //create subscriber
    ros::Subscriber MyMsgSub = nh.subscribe("MyMsg",1000,&MyMsgCallback);
    ros::spin();
    return 0;
}
```

类似之前创建的接收端程序，不过这里使用的是自定义的学生信息的数据类型。

说明如下。

- void MyMsgCallback(const book_message::Student& sdtInfo)：回调函数，对接收到的学生信息进行打印处理。
- ros::Subscriber MyMsgSub = nh.subscribe("MyMsg",1000,&MyMsgCallback)：创建接收者，并通过回调函数接收和处理 MyMsg 主题上的消息。

2.3.4 CMakeLists.txt文件

最终的 CMakeLists.txt 文件内容如下：

```
cmake_minimum_required(VERSION 2.8.3)
project(book_message)


find_package(catkin REQUIRED COMPONENTS
  message_generation roscpp rospy std_msgs)

add_message_files(FILES
  Student.msg
)

generate_messages(DEPENDENCIES
  std_msgs
)

catkin_package(
  CATKIN_DEPENDS
  message_runtime
)
```

```
include_directories(
  include ${catkin_INCLUDE_DIRS}
)

add_executable(Book_MyMsgPub
  src/Book_MyMsgPub.cpp
)
add_dependencies(Book_MyMsgPub ${${PROJECT_NAME}_EXPORTED_TARGETS} ${catkin_
EXPORTED_TARGETS})
target_link_libraries(Book_MyMsgPub
  ${catkin_LIBRARIES}
)

add_executable(Book_MyMsgSub
  src/Book_MyMsgSub.cpp
)
add_dependencies(Book_MyMsgSub ${${PROJECT_NAME}_EXPORTED_TARGETS} ${catkin_
EXPORTED_TARGETS})
target_link_libraries(Book_MyMsgSub
  ${catkin_LIBRARIES}
)
```

编译获得可执行文件 Book_MyMsgPub 和 Book_MyMsgSub。

2.3.5 测试

在两个终端分别输入如下内容：

```
rosrun book_message Book_MyMsgPub LiLei 180 160
rosrun book_message Book_MyMsgSub
```

可看到如下打印信息：

```
[ INFO] [1551684824.756760205]: The student information is:
name:LiLei---height:180.000000---weight:160.000000
[ INFO] [1551684824.856785386]: The student information is:
 name:LiLei---height:180.000000---weight:160.000000
[ INFO] [1551684824.956813035]: The student information is:
 name:LiLei---height:180.000000---weight:160.000000
```

在新终端运行 rosrun rqt_graph rqt_graph，查看节点关系图，如图 2-13 所示。

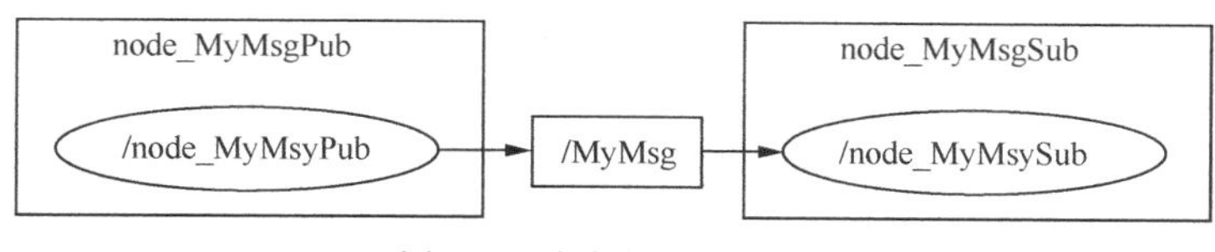

图 2-13 自定义消息节点图

2.4 服务

2.4.1 服务通信

在 ROS 中，除了消息这种通信类型外，还有一种称为服务的通信类型。不同于消息通信是单向的，服务是一种双向的通信，可以对接收到的请求做出回应。接下来练习如何使用自定义的消息类型进行通信测试，在本例中，创建一个服务文件，用于存储请求内容（输入密码）和返回的结果（密码是否正确）。

2.4.2 自定义srv

在 RoboWare Studio 中，通过 Add Srv Folder 创建 srv 文件夹，通过 Add Srv File 创建 PassWord.srv 文件，内容如下：

```
int64 password
---
bool result
```

该 srv 文件主要包含请求与响应部分。

- password 为 int64 类型的请求内容。
- result 是 bool 类型的响应。

类似自定义消息，在 CMakeLists.txt 文件中添加如下内容：

```
find_package(catkin REQUIRED COMPONENTS roscpp rospy std_msgs message_generation)
add_service_files(FILES
  PassWord.srv
)
generate_messages(DEPENDENCIES
```

```
  std_msgs
)
```

编译之后，在 .../robware_ws/devel/include/book_service 文件夹下可获得 3 个自动生成的文件：PassWord.h、PassWordRequest.h、PassWordResponse.h。

2.4.3 创建服务器

新建文件 ServerForBeginner.cpp，内容如下：

```
#include "ros/ros.h"
#include "book_service/PassWord.h"

// server callback function
bool serverCallback(book_service::PassWord::Request& req,
                    book_service::PassWord::Response& res)
{
    //if password = 123456, result is true, or result is false
    res.result = (req.password == 123456) ? true:false;
    return true;
}

int main(int argc,char** argv)
{
    //initial and name node
    ros::init(argc,argv,"server_node");
    //create nodehandle
    ros::NodeHandle nh;
    //create service server
    ros::ServiceServer serv = nh.advertiseService("pswserver",&serverCallback);
    ros::spin();
    return 0;
}
```

说明如下。

- bool serverCallback(book_service::PassWord::Request& req，book_service::PassWord::Response& res)：服务端回调函数，对客户端发出的请求做出相应的响应，在本例中会根据客户端发出的密码做出密码正确或错误的响应。
- ros::ServiceServer serv = nh.advertiseService("pswserver",&serverCallback)：创建服务器，并通过回调函数对请求做出响应。

2.4.4 创建客户端

新建文件 ClientForBeginner.cpp，内容如下：

```
#include "ros/ros.h"
#include "book_service/PassWord.h"
#include <cstdlib>

using namespace std;
int main(int argc, char** argv)
{
    //initial and name node
    ros::init(argc,argv,"node_client");
    if(argc<2)
    {
        cout<<"Error paramster,please run eg: rosrun book_service ClientForBeginner
123456"<<endl;
        return 1;
    }
    //create nodehandle
    ros::NodeHandle nh;
    //create client
    ros::ServiceClient client = nh.serviceClient<book_service::PassWord>("pswserver",
100);
    book_service::PassWord srv;
    //convert argv from char to int and pass value to service response
    srv.request.password = atoi(argv[1]);
    if(client.call(srv))
    {
        ROS_INFO("client connect success!");
        if(srv.response.result)
        {
            ROS_INFO("Welcom,correct password!");
        }else{
            ROS_INFO("Sorry,password error!");
        }
    }else{
        ROS_INFO("client connet fail!");
    }
    return 0;
}
```

说明如下。

- if(argc<2)：如果命令参数错误，提示用户如何正确填写命令行。
- ros::ServiceClient client = nh.serviceClient<book_service::PassWord>("pswserver",100)：创建客户端。
- book_service::PassWord srv：创建服务。
- srv.request.password = atoi(argv[1])：命令行第二个参数赋值给服务的请求。
- if(client.call(srv))：如果通信成功，打印 success，否则打印 fail。
- if(srv.response.result)：根据服务器响应打印密码正确或错误的信息。

2.4.5 CMakeLists.txt文件

在 CMakeLists.txt 文件中添加如下内容：

```
cmake_minimum_required(VERSION 2.8.3)
project(book_service)


find_package(catkin REQUIRED COMPONENTS roscpp rospy std_msgs message_generation)

add_service_files(FILES
  PassWord.srv
)

generate_messages(DEPENDENCIES
  std_msgs
)

catkin_package(
  CATKIN_DEPENDS
  message_runtime
)

include_directories(
  include ${catkin_INCLUDE_DIRS}
)

add_executable(ServerForBeginner
  src/ServerForBeginner.cpp
```

```
    )
    add_dependencies(ServerForBeginner ${${PROJECT_NAME}_EXPORTED_TARGETS} ${catkin_
EXPORTED_TARGETS})
    target_link_libraries(ServerForBeginner
      ${catkin_LIBRARIES}
    )

    add_executable(ClientForBeginner
      src/ClientForBeginner.cpp
    )
    add_dependencies(ClientForBeginner ${${PROJECT_NAME}_EXPORTED_TARGETS} ${catkin_
EXPORTED_TARGETS})
    target_link_libraries(ClientForBeginner
      ${catkin_LIBRARIES}
    )
```

编译获得可执行文件 ClientForBeginner 和 ServerForBeginner。

2.4.6 测试

在不同终端分别运行如下代码：

```
rosrun book_service ClientForBeginner 123456
rosrun book_service ServerForBeginner
```

可看到如下打印内容：

```
[ INFO] [1551686984.811029702]: client connect success!
[ INFO] [1551686984.811119481]: Welcom,correct password!
```

如果运行如下代码：

```
rosrun book_service ClientForBeginner
```

可看到如下提示信息：

```
Error paramster,please run eg: rosrun book_service ClientForBeginner 123456
```

若输入错误密码，如：

```
rosrun book_service ClientForBeginner 12345
```

可得到如下打印信息：

```
[ INFO] [1551687286.099570151]: client connect success!
[ INFO] [1551687286.099667373]: Sorry,password error!
```

2.5 参数

ROS 中通常包含许多参数，这些参数在可执行程序中起到了非常关键的作用。ROS 为我们提供了 3 种不同的方式来设置和获取参数，接下来将练习如何编写可执行文件，通过不同的方法来设置和获取参数。

2.5.1 编写参数设置获取节点

参数的设置和获取通常可以通过命令行实现，但是为了便于维护和排查错误，本节我们将通过编写可执行文件的方式来演示获取和设置参数的不同方法。ROS 中有如下 3 种获取参数的方式。

- ros::param::get()
- ros::NodeHandle::getParam()
- ros::NodeHandle::param()

需要注意，第三种 ros::NodeHandle::param() 方式在获取失败时会自动设置一个默认参数值。

设置参数的方式主要有以下两种。

- ros::param::set()
- ros::NodeHandle::setParam()

创建文件 book_param.cpp，内容如下：

```
#include "ros/ros.h"
#include <cstdlib>

using namespace std;

int main(int argc,char** argv)
{
    //initial and name node
    ros::init(argc,argv,"node_param");
    if(argc!=2)
    {
        cout<<"Error command paramter! Please run command eg:\n"\
            <<"rosrun book_param book_param 1\n"\
```

```
            <<"help information:\n"\
            <<" 1 ------ set param mode(ros::param::set())\n"\
            <<" 2 ------ set param mode(ros::NodeHandle::setParam())\n"\
            <<" 3 ------ get param mode(ros::param::get())\n"\
            <<" 4 ------ get param mode(ros::NodeHandle::getParam())\n"\
            <<" 5 ------ get param mode(ros::NodeHandle::param())\n"\
            <<endl;
        return 1;
    }

    //create node handle
    ros::NodeHandle nh;
    //param variable
    int IntParam;
    string StrParam;
    bool isIntParam, isStrParam;
    //mode flag
    int flag = atoi(argv[1]);

    // set or get param with different ways
    switch(flag)
    {
        case 1:
            ROS_INFO("set param mode(ros::param::set()):");
            ros::param::set("IntParam",1);
            ros::param::set("StrParam","stringdemo");
            break;
        case 2:
            ROS_INFO("set param mode(ros::NodeHandle::setParam()):");
            nh.setParam("IntParam",1);
            nh.setParam("StrParam","stringdemo");
            break;
        case 3:
            ROS_INFO("get param mode(ros::param::get()):");
            isIntParam = ros::param::get("IntParam",IntParam);
            isStrParam = ros::param::get("StrParam",StrParam);

            if(isIntParam){
                ROS_INFO("The IntParam is:%d",IntParam);
            }else{
                ROS_INFO("Get IntParam fail!");
            }

            if(isIntParam){
```

```
                ROS_INFO("The StrParam is:%s",StrParam.c_str());
            }else{
                ROS_INFO("Get StrParam fail!");
            }
            break;
        case 4:
            ROS_INFO("get param mode(ros::NodeHandle::getParam()):");
            isIntParam = nh.getParam("IntParam",IntParam);
            isStrParam = nh.getParam("StrParam",StrParam);

            if(isIntParam){
                ROS_INFO("The IntParam is:%d",IntParam);
            }else{
                ROS_INFO("Get IntParam fail!");
            }

            if(isIntParam){
                ROS_INFO("The StrParam is:%s",StrParam.c_str());
            }else{
                ROS_INFO("Get StrParam fail!");
            }
            break;
        case 5:
            ROS_INFO("get param mode(ros::NodeHandle::param()):");
            //warning: this way will set default value when get no param!
            nh.param("IntParam",IntParam,11);
            // be careful when use ros::NodeHandle::param get string param!
            nh.param<std::string>("StrParam",StrParam,"stringdemo_default");
            ROS_INFO("\nThe IntParam is:%d\nThe StrParam is:%s",IntParam,StrParam.
c_str());
            break;
        default:
            ROS_INFO("flag value is not in range: [1,5]");
    }
    return 0;
}
```

说明如下。

- if(argc!=2)：如果用户输入命令行错误，则打印提示信息，有 5 种模式展示如何获取和设置参数。

```
int IntParam;
string StrParam;
```

分别用于设置 int 和 string 两种类型的参数。

- int flag = atoi(argv[1])：模式选择标志位。
- case 1：ros::param::set() 方法设置 IntParam 和 StrParam 参数。
- case 2：ros::NodeHandle::setParam() 方法设置 IntParam 和 StrParam 参数。
- case 3：ros::param::get() 方法获取 IntParam 和 StrParam 参数。
- case 4：ros::NodeHandle::getParam() 方法获取 IntParam 和 StrParam 参数。
- case 5：ros::NodeHandle::param() 方法获取 IntParam 和 StrParam 参数。

2.5.2 CMakeLists.txt文件

创建 CMakeLists.txt 文件，内容如下：

```
cmake_minimum_required(VERSION 2.8.3)
project(book_param)

find_package(catkin REQUIRED COMPONENTS roscpp rospy std_msgs)

catkin_package(
)

include_directories(
  include ${catkin_INCLUDE_DIRS}
)

add_executable(book_param
  src/book_param.cpp
)
add_dependencies(book_param ${${PROJECT_NAME}_EXPORTED_TARGETS} ${catkin_
EXPORTED_TARGETS})
target_link_libraries(book_param
  ${catkin_LIBRARIES}
)
```

编译后获取 book_param 可执行文件。

2.5.3 测试

打开终端，运行：

```
rosrun book_param book_param
```

打印错误提示信息如下：

```
Error command paramter! Please run command eg:
rosrun book_param book_param
help information:
 1 ------ set param mode(ros::param::set())
 2 ------ set param mode(ros::NodeHandle::setParam())
 3 ------ get param mode(ros::param::get())
 4 ------ get param mode(ros::NodeHandle::getParam())
 5 ------ get param mode(ros::NodeHandle::param())
```

若运行参数不在设定范围内，如：

```
rosrun book_param book_param 6
```

则打印出的错误提示信息如下：

```
[ INFO] [1551689308.305949069]: flag value is not in range: [1,5]
```

重新运行：

```
rosrun book_param book_param 1
```

打印提示信息：

```
[ INFO] [1551688969.788377908]: set param mode(ros::param::set()):
```

查看参数列表：

```
rosparam list
```

打印出所有参数：

```
/IntParam
/StrParam
```

命令行获取参数值：

```
rosparam get StrParam
```

打印出参数值：

```
Stringdemo
```

你也可以分别运行 rosrun book_param book_param 2(3,4,5) 来测试不同模式设置和获取参数的效果。

2.6 动态参数设置

通常，调试时（尤其是在导航及建图应用中）需要经常修改程序中的参数值，这时无论是修改命令行，还是编写固定修改参数的可执行文件，都无法满足要求。ROS 为我们提供了动态参数设置的机制，接下来我们将练习编写具备动态参数设置功能的可执行文件。

2.6.1 创建cfg文件

创建动态参数 ROS 包 book_dynamic_param，加入依赖项 roscpp,rospy,dynamic_reconfigure。在功能包下新建 cfg 文件夹，并创建 DynamicParam.cfg 文件，内容如下：

```
#!/usr/bin/env python
PACKAGE = "book_dynamic_param"

from dynamic_reconfigure.parameter_generator_catkin import *

gen = ParameterGenerator()

gen.add("IntDyParam",int_t,0,"An Int Parameter",0,0,9)
gen.add("DouDyParam",double_t,0,"A Double Parameter",1.5,0,9)
gen.add("StrDyParam",str_t,0,"A String Parameter","Hello,I'm Robot!")
gen.add("BoolDyParam",bool_t,0,"A Bool Parameter",True)

student_info = gen.enum([gen.const("Name",str_t,"LiLei","Name Information"),
                         gen.const("Sex",str_t,"Man","Sex Information"),
                         gen.const("Age",str_t,"18","Age Information")],
                        "A set contain a student information")

gen.add("StudentInfo",str_t,0,"A studenet information set","LiLei",edit_
method=student_info)
exit(gen.generate(PACKAGE,"node_DynamicParam","DynamicParam"))
```

该配置文件使用 Python 语言实现，首先需要导入 dynamic_reconfigure 功能包提供的参数生成器，通过 gen = ParameterGenerator() 创建生成器。这里定义了 4 个不同类型的参数：int_t、double_t、str_t、bool_t。使用参数生成器的 add(name,type,level,description, default, min,max) 方法生成参数，具体用法如下。

- name：参数名，使用字符串描述。
- type：定义参数的类型，可以是 int_t、double_t、str_t 或者 bool_t。
- level：需要传入参数动态配置回调函数中的掩码，在回调函数中会修改所有参数的掩码，表示参数已经进行修改。
- description：描述参数作用的字符串。
- default：设置参数的默认值。
- min：可选，设置参数的最小值，对于字符串和布尔类型值不生效。
- max：可选，设置参数的最大值，对于字符串和布尔类型值不生效。

然后利用 gen.enum 方法生成一个枚举类型的值，最后通过 exit 生成所有与 C++ 和 Python 相关的文件并退出程序，这里第二个参数表示动态参数运行的节点名，第三个参数是生成文件所使用的前缀，需要和配置文件 DynamicParam.cfg 名称相同。

配置文件创建完成后，需要为配置文件添加可执行权限，命令如下：

```
chmod a+x cfg/DynamicParam.cfg
```

在 CMakeLists.txt 文件中添加以下内容：

```
generate_dynamic_reconfigure_options(
    cfg/DynamicParam.cfg
)
add_dependencies(hell ${PROJECT_NAME}_gencfg)
```

> **注意**
>
> hell 是工作空间中已编译好的可执行节点名，读者可根据自己工作空间中的节点任意替换（如前文编译的 PubForBeginner）。

编译之后，在工作空间目录 .../devel/include/book_dynamic_param 文件夹下可得到自动生成的 DynamicParamConfig.h 头文件。

2.6.2 创建动态参数设置可执行文件

创建 book_dyparam.cpp 文件，并添加如下内容：

```
#include "ros/ros.h"
#include "dynamic_reconfigure/server.h"
#include "book_dynamic_param/DynamicParamConfig.h"
```

```
//define call back function
void paramCallback(book_dynamic_param::DynamicParamConfig& config,uint32_t level)
{
    ROS_INFO("Request: %d %f %s %s %s",
                config.IntDyParam,config.DouDyParam,
                config.StrDyParam.c_str(),
                config.BoolDyParam?"True":"False",
                config.StudentInfo.c_str());
}

int main(int argc, char** argv)
{
    //initial and name node
    ros::init(argc,argv,"node_DynamicParam");
    //create node handle
    dynamic_reconfigure::Server<book_dynamic_param::DynamicParamConfig> server;
    dynamic_reconfigure::Server<book_dynamic_param::DynamicParamConfig>::CallbackType f;

    f = boost::bind(&paramCallback,_1,_2);
    server.setCallback(f);
    ros::spin();
    return 0;
}
```

说明如下。

- dynamic_reconfigure::Server<book_dynamic_param::DynamicParamConfig> server：创建了一个参数动态配置的服务端实例，参数配置的类型就是配置文件中描述的类型。该服务端实例会监听客户端的参数配置请求。
- dynamic_reconfigure::Server<book_dynamic_param::DynamicParamConfig>::CallbackType f：定义回调函数，并将回调函数和服务端绑定，当客户端请求修改参数时，服务端即可跳转到回调函数中进行处理。回调函数有两个传入参数，一个是新的参数配置值，另一个是表示参数修改的掩码。然后，通过 ROS_INFO 在回调函数中将修改后的参数值打印出来。

2.6.3 CMakeLists.txt文件

CMakeLists.txt 文件内容如下：

```
cmake_minimum_required(VERSION 2.8.3)
project(book_dynamic_param)
```

```
find_package(catkin REQUIRED COMPONENTS roscpp rospy std_msgs dynamic_reconfigure)
generate_dynamic_reconfigure_options(
    cfg/DynamicParam.cfg
)
catkin_package(
)
include_directories(
  include ${catkin_INCLUDE_DIRS}
add_dependencies(hell ${PROJECT_NAME}_gencfg)

add_executable(book_dyparam
  src/book_dyparam.cpp
)
add_dependencies(book_dyparam ${${PROJECT_NAME}_EXPORTED_TARGETS} ${catkin_
EXPORTED_TARGETS})
target_link_libraries(book_dyparam
  ${catkin_LIBRARIES}
)
```

2.6.4 测试

编译后获得可执行文件，运行如下命令：

```
rosrun book_dynamic_param book_dyparam
rosrun rqt_reconfigure rqt_reconfigure
```

可看到可视化调试界面如图 2-14 所示。

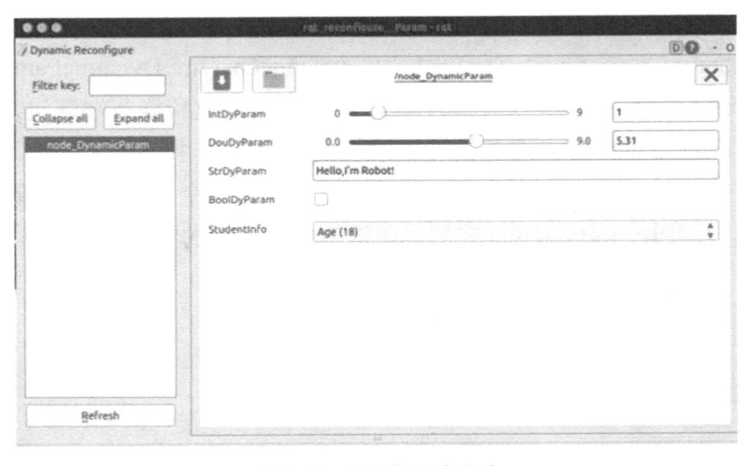

图 2-14　可视化调试界面

滑动滑柄并勾选不同参数，可看到类似如下内容的打印信息：

```
[ INFO] [1553048397.886199339]: Request: 1 1.500000 Hello,I'm Robot! True LiLei
[ INFO] [1553048410.264597290]: Request: 1 1.500000 Hello,I'm Robot! True Man
[ INFO] [1553048419.952933985]: Request: 1 1.500000 Hello,I'm Robot! True 18
[ INFO] [1553048421.895419021]: Request: 1 1.500000 Hello,I'm Robot! False 18
[ INFO] [1553048428.313038319]: Request: 1 5.310000 Hello,I'm Robot! False 18
```

查看节点运行关系图，如图 2-15 所示。

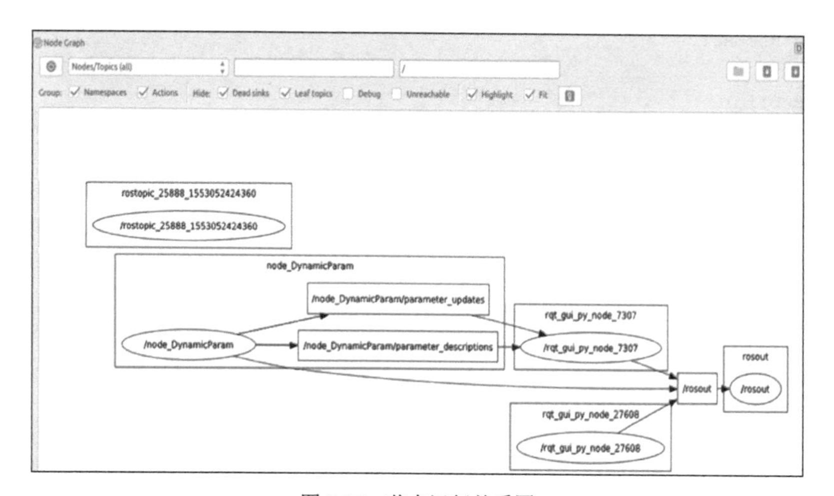

图 2-15　节点运行关系图

2.7 ROS类编程思想

使用 C++ 语言编程的读者应该明白，通过类可以极大地提高代码的复用性和开发效率。在 ROS 中，我也特别推荐读者能够养成使用类进行开发的习惯，这样可以有效地管理开发程序，尤其是大规模变量的使用与控制。至此，本书已经讲解了有关 ROS 开发的大部分基础知识，能够满足大部分读者常用的开发需求。本节，我们将学习使用类对已学过的知识进行编程开发。

2.7.1 创建类头文件

创建 book_class.h 头文件，文件中的内容如下：

```
#include "ros/ros.h"
#include "std_msgs/String.h"
#include "std_msgs/Float64.h"
#include "std_msgs/Int32.h"
#include "book_class/PassWord.h"
#include <cstdlib>

using namespace std;

class RosBasics
{
    public:
        RosBasics();
        RosBasics(ros::NodeHandle& nh);
        ~RosBasics();
        void initPublish();
        void initSubscribe();
        void initServer();
        bool serverCallback(book_class::PassWord::Request& req,
                            book_class::PassWord::Response& res);
        void msgCallback(const std_msgs::Float64& stdmsg);
        void paramTest();
    private:
        ros::NodeHandle nh_;
        ros::Publisher pub_;
        ros::Subscriber sub_;
        ros::ServiceServer serv_;
};
```

说明如下。

- #include "book_class/PassWord.h" 中的头文件类似之前创建的 srv，内容也相同，此处不再详细介绍其创建过程。读者可参考 3.4.2 节中自定义 srv 的过程，自行创建。
- class RosBasics 定义了一个 RosBasics 类，其构造函数和析构函数为：

```
RosBasics();
RosBasics(ros::NodeHandle& nh);
~RosBasics();
```

- Public 部分还包含了一些发布者、订阅者、服务的初始化、消息回调函数、服务回调函数、参数设置函数的定义等。私有属性则包含了 ROS 句柄、Publisher、Subscriber、ServiceServer 等。

2.7.2 创建类应用可执行文件

创建 book_class.cpp 文件，实现头文件中的函数并进行程序功能设计，文件中的内容如下：

```
#include "book_class.h"
RosBasics::RosBasics():nh_("~")
{
    initPublish();
    initServer();
    initSubscribe();
    paramTest();
}

RosBasics::RosBasics(ros::NodeHandle& nh):nh_(nh)
{
    initPublish();
    initServer();
    initSubscribe();
    paramTest();
}

RosBasics::~RosBasics()
{
}

void RosBasics::initPublish()
{
    ROS_INFO("publish initializing!");
    pub_ = nh_.advertise<std_msgs::Float64>("Topic2",100);
}

void RosBasics::initSubscribe()
{
    ROS_INFO("subscribe initializing!");
    sub_ = nh_.subscribe("Topic1",1,&RosBasics::msgCallback,this);
}

void RosBasics::initServer()
{
    ROS_INFO("server initializing!");
```

```
    serv_ = nh_.advertiseService("pwdcheck",&RosBasics::serverCallback,this);
}

bool RosBasics::serverCallback(book_class::PassWord::Request& req,
                               book_class::PassWord::Response& res)
{
    res.result = (req.password == 123456) ? true:false;
    if(res.result)
    {
        ROS_INFO("Welcom, password correct!");
    }else{
        ROS_INFO("Sorry, password error!");
    }
    return true;
}

//obtain message from Topic1 and add value then publish to Topic2
void RosBasics::msgCallback(const std_msgs::Float64& stdmsg)
{
    std_msgs::Float64 msgpub;
    msgpub.data = stdmsg.data + 2;
    ROS_INFO("Receive date from Topic1 content is:%f",stdmsg.data);
    ROS_INFO("Publish date with Topic2 content is:%f",msgpub.data);
    pub_.publish(msgpub);
}

//test ros param
void RosBasics::paramTest()
{
    int Age;
    bool checkAge;
    checkAge = nh_.getParam("Age",Age);
    if(checkAge)
    {
        ROS_INFO("Welcome,Get Param Success!");
        if(Age>=18)
        {
            ROS_INFO("You are adult!");
        }else{
            ROS_INFO("You are children!");
        }
    }else{
        ROS_INFO("Sorry,Get Param Fail!");
    }
```

```
}

int main(int argc,char** argv)
{
    //initial and name node
    ros::init(argc,argv,"node_classdemo");
    //create node handle
    ros::NodeHandle nh;
    //instantiate class object
    RosBasics classdemo(nh);
    ros::spin();
    return 0;
}
```

说明如下。

首先在构造函数中为参数初始化赋值，然后将需要执行的程序按照相应的顺序放进构造函数。

- void RosBasics::initPublish()：在“Topic2”主题上发布 std_msgs::Float64 类型的消息。
- void RosBasics::initSubscribe()：订阅“Topic1”主题上的消息。
- void RosBasics::initServer()：服务端初始化。
- bool RosBasics::serverCallback(book_class::PassWord::Request& req,book_class::PassWord::Response& res)：服务回调函数，对接收的请求做出相应的响应。在本例中，如果收到的请求与正确密码 123456 相同，则打印“Welcom, password correct!”，否则打印“Sorry, password error!”。
- void RosBasics::msgCallback(const std_msgs::Float64& stdmsg)：消息回调函数，在本例中，对“Topic1”主题上接收到 std_msgs::Float64 的消息数值进行增加 2 操作，然后将值增加后的消息发布到“Topic2”主题上。
- void RosBasics::paramTest()：参数测试函数，主要功能为对获取的年龄参数进行检测。若年龄 Age>=18，打印“You are adult!”，否则打印“You are children!”。

在主函数中创建 ros::NodeHandle 句柄，并实例化 RosBasics 类。

2.7.3 CMakeLists.txt文件

创建 CMakeLists.txt 文件，内容如下：

```
cmake_minimum_required(VERSION 2.8.3)
project(book_class)
find_package(catkin REQUIRED COMPONENTS
  message_generation roscpp rospy std_msgs dynamic_reconfigure)
add_service_files(FILES
  PassWord.srv
)

generate_messages(DEPENDENCIES
  std_msgs
)
catkin_package(
  CATKIN_DEPENDS
  message_runtime
)
include_directories(
  include ${catkin_INCLUDE_DIRS}
# include
# ${catkin_INCLUDE_DIRS}
)
add_executable(book_class
  src/book_class.cpp
)
add_dependencies(book_class ${${PROJECT_NAME}_EXPORTED_TARGETS} ${catkin_
EXPORTED_TARGETS})
target_link_libraries(book_class
  ${catkin_LIBRARIES}
)
```

2.7.4 测试

编译获取可执行文件，打开终端，运行：

```
rosrun book_class book_class
```

可获得如下提示信息：

```
[ INFO] [1553051662.685960008]: publish initializing!
[ INFO] [1553051662.688131239]: server initializing!
[ INFO] [1553051662.690090038]: subscribe initializing!
[ INFO] [1553051662.700113803]: Sorry,Get Param Fail!
```

新开终端，运行：

```
rostopic pub -r 10 /Topic1 std_msgs/Float64 1
```

打印提示信息如下：

```
[ INFO] [1553051775.158305170]: Receive date from Topic1 content is:1.000000
[ INFO] [1553051775.158419240]: Publish date with Topic2 content is:3.000000
```

在新终端中运行：

```
rosservice call pwdcheck 123456
```

打印提示信息如下：

```
[ INFO] [1553051893.497096369]: Welcom, password correct!
result: True
```

测试参数服务，运行下面的指令：

```
rosparam set Age 18
```

关闭 book_class 节点并重新运行，得到打印提示信息如下：

```
[ INFO] [1553052136.439965048]: Welcome,Get Param Success!
[ INFO] [1553052136.440955370]: You are adult!
```

运行 rosrun rqt_graph rqt_graph，查看节点关系图，如图 2-16 所示。

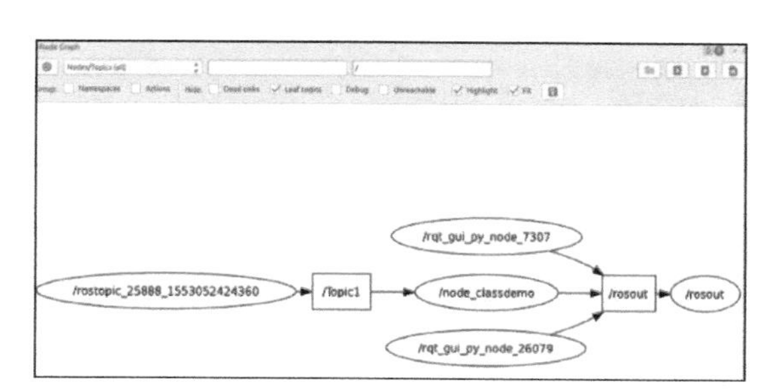

图 2-16　节点关系图

第3章 调试及仿真工具

ROS 自带丰富的可视化调试和仿真工具，能够极大地提高开发效率。例如，最基本的可视化调试工具 Rviz，用于机器人仿真环境搭建的 Gazebo；为 ROS 准备的基于 Qt 框架的 GUI 开发环境 rqt（包含 rqt、rqt_common_plugins 和 rqt_robot_plugins）；录制和回放数据的 rosbag；用于非 ROS 系统和 ROS 系统进行交互通信的 rosbridge 功能包等。本章将围绕以上几种工具及开发工作进行讲解。

3.1 Rviz

Rviz（ros-visualization）是 ROS 官方提供的一款三维可视化工具，我们需要用到的所有机器人相关数据几乎全部可以在 Rviz 中展现。作为一种可扩展化的视图工具，Rviz 可以使用插件机制来扩展丰富的功能，进行二次开发。Rviz 中经常使用的激光数据可视化显示、图像数据可视化显示功能，其实都是官方提供的插件，用户也可以根据实际开发需求来开发自己的插件。

启动 Rviz，终端运行命令：

```
rosrun rviz rviz
```

可看到图 3-1 所示的 Rviz 界面。

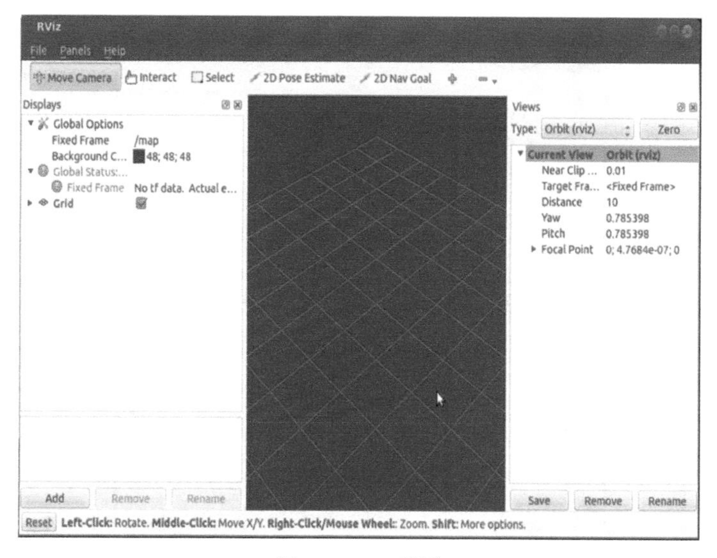

图 3-1　Rviz 界面

在图 3-1 中，界面中间是三维网格可视化窗口，左边的显示列表（Displays）列出了用户添加的显示项，右边则包含一些全视角和坐标系的选项。

若要添加显示项（Adding a new display），可按组合键 <Ctrl+N> 或者单击图 3-2 中的“Add”按钮。

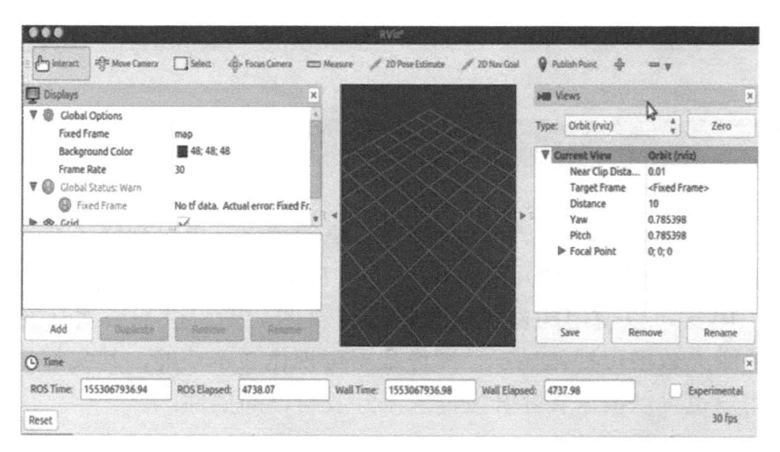

图 3-2　添加显示项

在弹出的窗口中，可供选择的显示插件包含坐标轴、相机、激光数据、点云数据等，具体内容如图 3-3 所示。

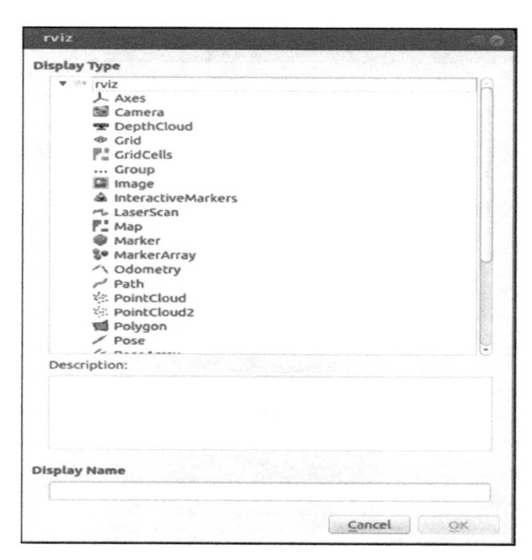

图 3-3　显示插件

每个显示项都有自己的具体属性，如在进行 slam 建图时，获取的激光雷达显示项的具体参数如图 3-4 所示。

从图 3-4 中可以看到雷达信息的 Topic 名称、显示形状、颜色、像素大小等信息。

显示项的状态用于帮助开发者查看该项是否正常运行，每个显示项都有 4 种状态，分别为 OK、Warning、Error 和 Disabled，4 种状态通过不同的颜色显示进行区分。如图 3-5 所示，分别用黄色、红色、灰色等表示不同的显示状态。

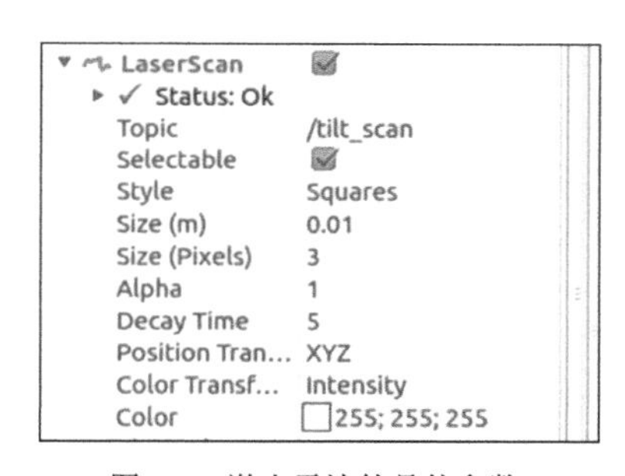

图 3-4　激光雷达的具体参数

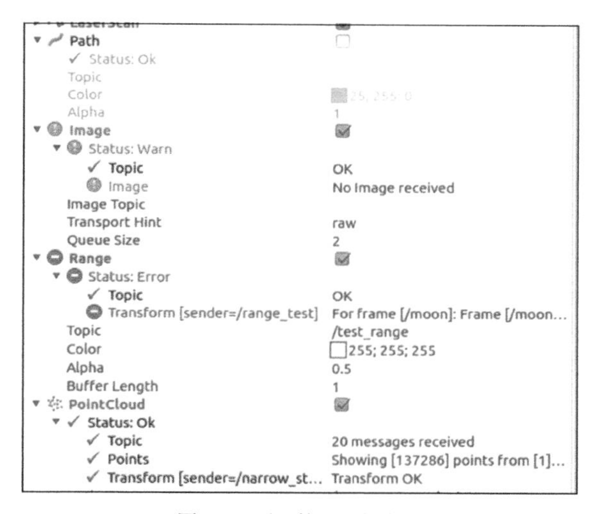

图 3-5　不同的显示状态

为了便于调试，通常需要将显示项进行重命名，以便开发者理解其含义。重命名方式有以下两种。

- 选中显示项，按组合键 <Ctrl+R> 进行重命名操作。
- 选中显示项，通过单击“Rename”按钮进行重命名。

在开发具体项目时，开发者往往需要根据具体需求配置不同的显示属性。例如，为 PR2 机器人配置的显示属性就不能使用小车的可视化操作。因此，可以保存不同的配置文件，在使用时根据具体项目需要加载不同的显示配置。

通常，显示配置文件包含以下 4 个部分。

- Displays and their properties
- Tools and their properties
- View controller type and settings for the initial viewpoint, plus saved views

- Window layout and the list of panels

在运行 rosrun rviz rviz 指令时，我们加载的是默认配置文件。改变不同的属性，配置文件将被重写，可通过 Save Config 或者按组合键 <Ctrl+S> 保存配置，也可通过 Save Config As 进行配置文件重命名。保存的 Rviz 配置文件在 ~/.rviz/ 目录下，可通过按组合键 <Ctrl+H> 查看隐藏的文件或文件夹。系统默认的文件为 ~/.rviz/default.rviz。

视角控制选项如图 3-6 所示，可通过 View 对视角和固定坐标轴进行设置。

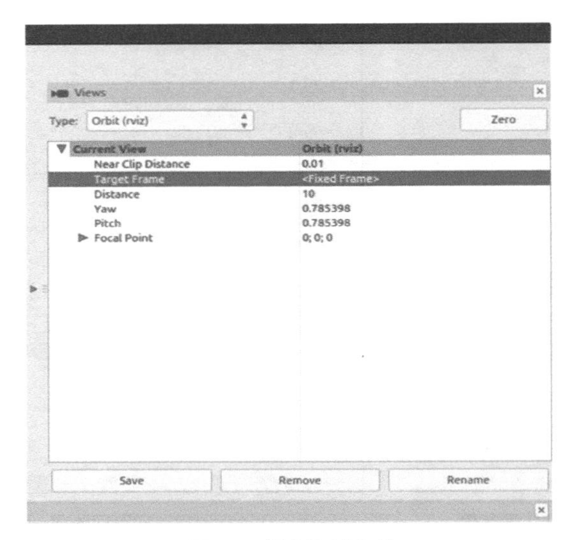

图 3-6 视角控制选项

在 Rviz 中，内置有不同的视角类型，如 Orbit、XYOrbit、FPS（first-person shooter）。默认是 Orbit 类型，可通过 Type 进行选择。

关于坐标系，ROS 使用 TF 转换系统发布不同坐标系之间的变换关系。在所有的坐标系中，最重要的是固定坐标系 Fixed Frame。固定坐标系是描述世界的参考坐标系，其他坐标系都是相对固定坐标系而言的，固定坐标系通常为“map”“world”或者“odometry frame”。

3.2 Gazebo

Gazebo 是一个三维动态模拟器，能够准确有效地模拟复杂的室内和室外环境中机器人的运动。与游戏引擎类似，Gazebo 为用户和程序提供了更高逼真度的物理模拟、传感器和接口。不同于二维平面的模拟器 Stage，Gazebo 是三维的模拟器，用户可以自己在地图上添加几何体。如图 3-7 所示，分别为 Stage 和 Gazebo 仿真环境。

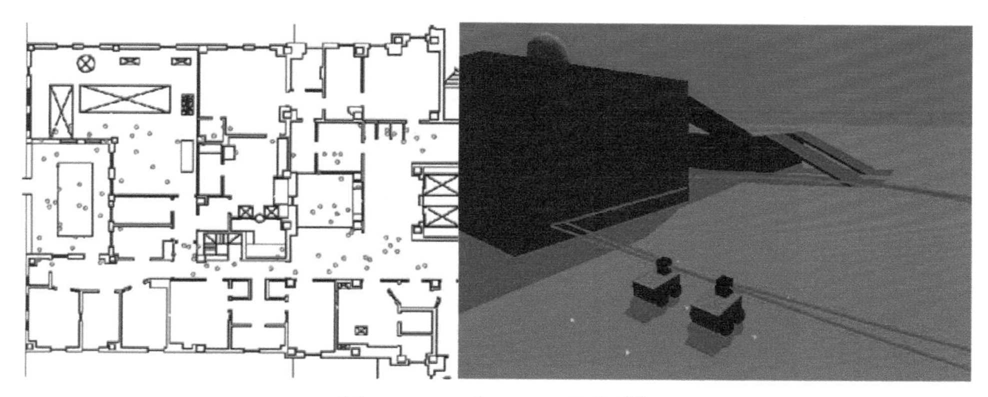

图 3-7　Stage 和 Gazebo 仿真环境

3.2.1　安装与更新

在安装 ROS 时，使用的命令为 ros-indigo-desktop-full，默认已经安装 Gazebo，无须重新安装。

若部分安装 ROS，则需使用如下命令进行安装：

```
sudo apt-get install ros-indigo-simulators
```

> **注意**
>
> 上述命令是针对 indigo 版本的 ROS 而言，读者需根据自己的 ROS 版本替换 indigo。

若使用的命令为 ros-indigo-desktop-ful，默认安装的 gazebo 版本是 2.2。若要更新版本，可通过以下命令（以更新 7.0 版本为例）进行安装：

```
$ sudo sh -c 'echo "deb http://packages.osrfoundation.org/gazebo/ubuntu-
stable `lsb_release -cs` main" > /etc/apt/sources.list.d/gazebo-stable.list'
$ wget http://packages.osrfoundation.org/gazebo.key -O - | sudo apt-key add -
$ sudo apt-get update
$ sudo apt-get install gazebo7
```

3.2.2 Gazebo环境

打开终端，输入命令 Gazebo，即可看到图 3-8 所示的软件环境。

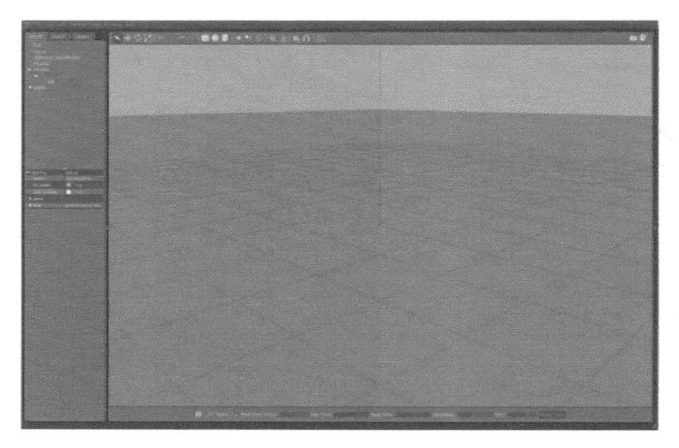

图 3-8 Gazebo 软件环境

界面中间栅格部分为场景，作为模拟器的主要组成部分，场景是模拟对象被放置的地方，也是用户与模拟器交互的可视化区域，如图 3-9 所示。

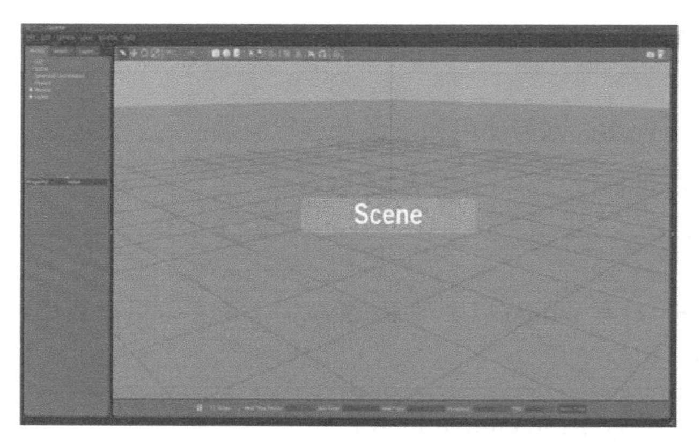

图 3-9 场景界面

通过拖动左右两侧面板的控制条，可以显示、隐藏面板或调整面板大小。在启动 Gazebo 时，默认显示左侧面板隐藏右侧面板。如图 3-10 所示，右侧面板主要是在场景中当物体被选中时才会弹出来，用于设置被选对象的属性；左侧面板主要由以下 3 个部分构成。

- World："世界"选项卡显示当前在场景中的模型，并允许查看和修改模型参数，比如它们的姿势。还可以通过展开"GUI"选项调整相机姿势来更改相机视图角度。
- Insert：向模拟添加新对象（模型）的位置插入选项卡。要查看模型列表，需要单击箭头小图标以展开文件夹，在要插入的模型上单击并释放，然后在场景中再次单击以添加它。
- Layers："层"选项卡组织和显示模拟中可用的不同可视化组。一个层可以包含一个或多个模型，打开或关闭图层将显示或隐藏该图层中的模型。

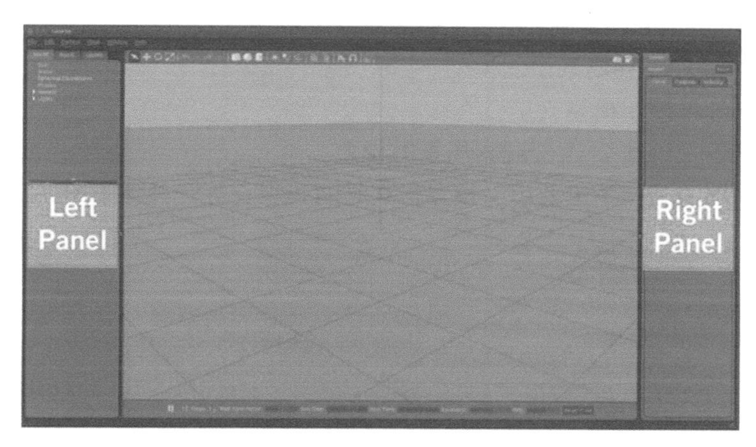

图 3-10　控制面板

3.2.3　选项卡与工具条

Gazebo 主要有两个工具条，分别位于场景的顶部和底部。场景顶部的工具条具体操作功能如图 3-11 所示。

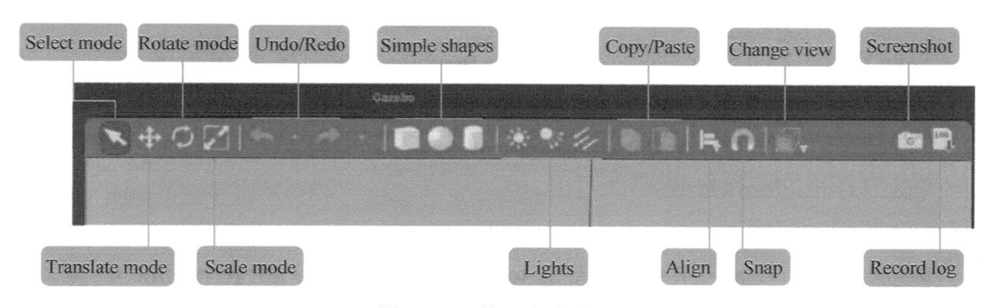

图 3-11　顶部工具条选项

各个操作的具体含义如下。

- Select mode：选择模式，用于选择操作对象。
- Translate mode：选中物体并移动。
- Rotate mode：选中物体并旋转。
- Scale mode：选中物体进行缩放。
- Undo/Redo：撤销 / 重做操作。
- Simple shapes：一些可供拖进场景的内置简单几何体。
- Lights：对场景添加灯光操作。
- Copy/Paste：复制 / 粘贴选中的对象。
- Align：对齐场景中的模型。
- Snap：捕获模型。
- Change view：改变观察场景的视角。

场景底部的工具条主要用于设置仿真时间和步长等，具体如图 3-12 所示。

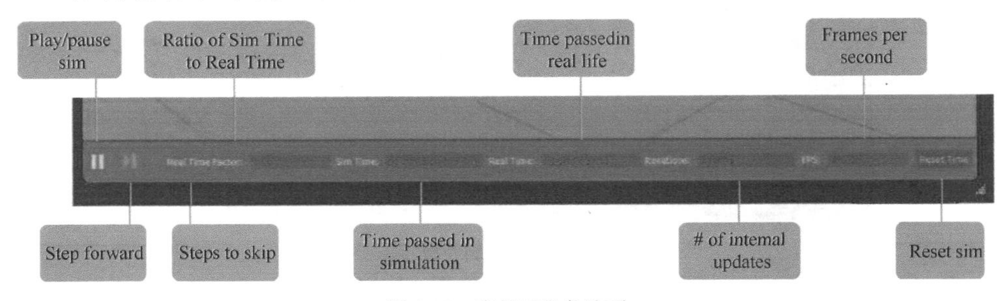

图 3-12 底部工具条选项

和大多数应用一样，Gazebo 的最上方是菜单栏。菜单栏主要包含 6 个部分：File、Edit、Camera、View、Window 和 Help。每个部分具体操作说明如下。

（1）File（文件）

Save World（Save the world as-is）：保存环境。

Save World As（Save the world as-is & provide a name）：环境另存为。

Save Configuration（Save your Gazebo interface configuration）：保存配置。

Clone world（Start new simulation from the current state）：复制环境。

Quit（QuitJclose Gazebo）：退出。

（2）Edit（编辑）

Reset Model Poses（Reset models to original poses; do not reset time）：重复模型位姿。

Reset World（Reset everything in world; reset time）：重置环境。

Building Editor（Construct a building）：构建编辑器。

Model Editor（Build an articulated object）：模型编辑器。

（3）Camera（相机）

Ormographic（View the scene with no perspective）：无透视情况下查看场景。

Perspective（View scene with perspective projection）：透视情况下查看场景。

FPS View Control（Control view as First-person Shooter: WASD keys & mouse）：第一视角。

Orbit View Control（Control view by rotating around a point with the mouse）：旋转视角。

Reset View Angle（Move view to the initial）：重置视角。

（4）View（显示）

Grid（Display ground plane grid）：栅格。

Origin（Display world origin）：原点。

Transparent（Display models as transparent）：透明。

Wireframe（Display models as wireframe）：显示线框。

Collisions（Display model collisions）：显示碰撞。

Joints（Display model joints）：显示关节。

Center of Mass（Display model centers of mass）：显示质心。

Inertias（Display model inertias）：显示模型惯性。

Contacts（Display collision points）：显示碰撞点。

Link Frames（Display coordinate frames for links）：显示链接的坐标系。

（5）Window（窗口）

Topic Visualization（Visualize messages being published per topic）：话题可视化。

Oculus Rift（Integrate an Oculus Rift）：虚拟现实。

Show GUI Overlays（Show overlays from GUI Plugins）：界面插件显示。

Show Toobars（Show/hide top & bottom toolbars）：工具显示。

Full Screen（Make the Scene full screen）：全屏。

（6）Help（帮助）

Hotkey Chart（Open hot key chart）：快捷键表。

About（View info about Gazebo）：相关信息。

使用 Gazebo 进行物体运动模拟时，需要选中模型并修改参数等，因此需要频繁地使用鼠标操作。Gazebo 中对应的鼠标操作说明如图 3-13 所示。

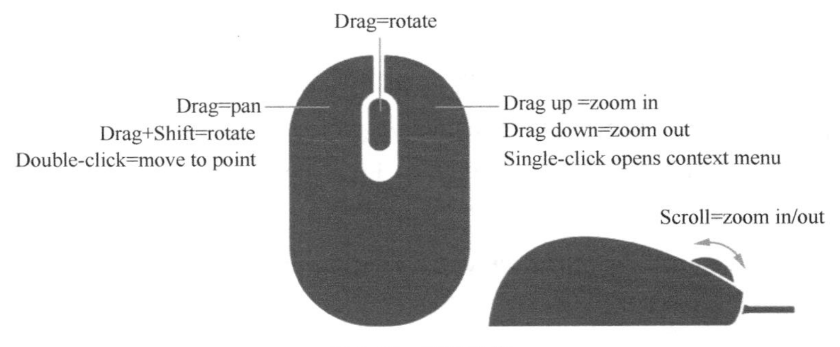

图 3-13 鼠标操作

3.2.4 模拟场景组成元素

模拟场景通常被称为世界，开发者可根据自己的需求设计不同属性的世界环境。世界通常包含的组成元素有 World、Models、Links、Collision Objects、Visual Objects、Sensors 和 Plugins，其组织结构如图 3-14 所示。

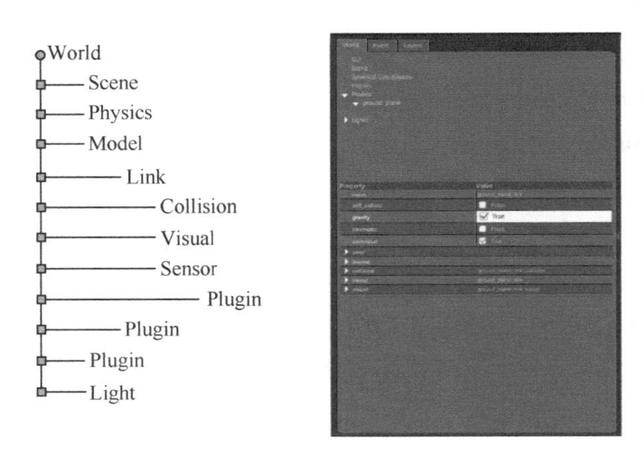

图 3-14 组织结构

世界文件以 .world 后缀结尾。通常，一个大型的世界文件由不同的小模型组成，这些模型文件通常由 SDF 或 URDF 文件编写，可被 Gazebo 直接加载。Willow Garage World 是 Gazebo 中自带的世界文件，其显示效果如图 3-15 所示。

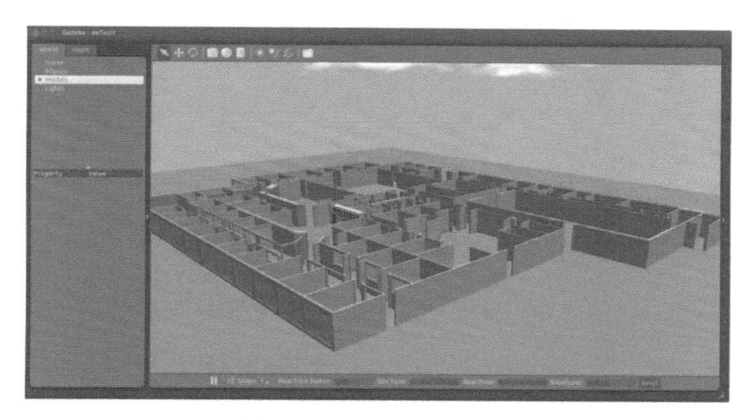

图 3-15 Willow Garage World

模型是场景文件的重要组成。一个模型通常包含几何形状、材质、颜色纹理、速度等物理属性。图 3-16 所示为 Turtlebot 在 Gazebo 中的模型。

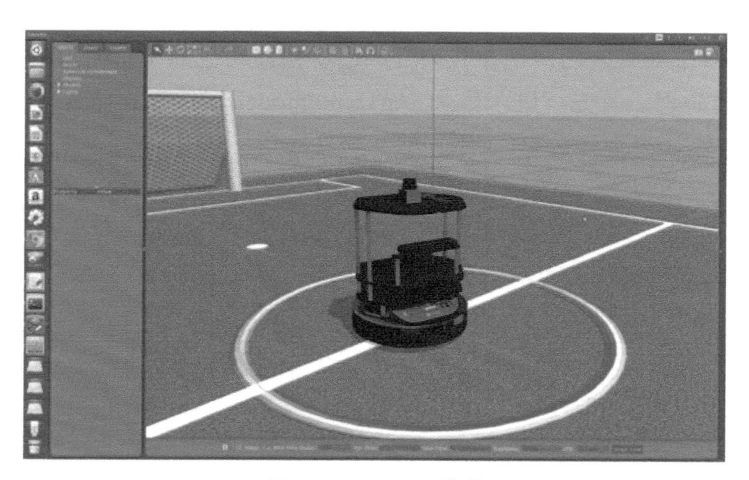

图 3-16 Turtlebot 模型

每个模型通常由不同的关节连接而成，在模型物理结构的基础上还会添加激光、视觉等传感器。在图 3-16 中，Turtlebot 机器人的顶部安装有激光雷达传感器。图 3-17 所示为 Gazebo 中激光雷达数据的可视化效果。

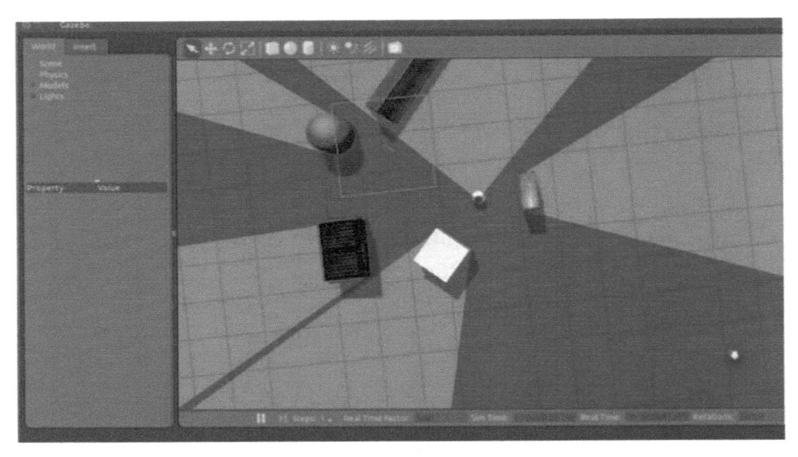

图 3-17　激光雷达数据可视化

3.2.5　搭建简单机器人模型

了解了基础知识之后，现在我们开始动手搭建自己的小车模型，并为其搭载 kinect 视觉传感器。首先，进入目录 ~/.gazebo/models，并创建 amy_robot 文件夹。

```
$ cd ~/.gazebo/models
$ mkdir amy_robot
```

在 amy_robot 文件夹下分别创建 model.config 和 model.sdf 文件。在 model.config 文件中添加如下内容：

```
<?xml version="1.0"?>
<model>
  <name>My Robot</name>
  <version>1.0</version>
  <sdf version='1.4'>model.sdf</sdf>

  <author>
   <name>My Name</name>
   <email>me@my.email</email>
  </author>

  <description>
    My awesome robot.
```

```
  </description>
</model>
```

说明如下。

- <model>：name 定义模型的名称和 sdf 文件的名称与版本。
- <author>：描述作者信息。
- <description>：描述模型的信息。

然后在 model.sdf 中创建小车的车体结构，添加车体：

```
<?xml version='1.0'?>
<sdf version='1.4'>
  <model name="my_robot">
      <static>true</static>
      <link name='chassis'>
        <pose>0 0 .1 0 0 0</pose>

        <collision name='collision'>
          <geometry>
            <box>
              <size>.4 .3 .1</size>
            </box>
          </geometry>
        </collision>

        <visual name='visual'>
          <geometry>
            <box>
              <size>.4 .3 .1</size>
            </box>
          </geometry>
        </visual>
```

该段主要定义车体的长、宽、高分别为 0.4、0.3、0.1。车体位姿坐标（0 0 0.1 0 0 0），前 3 个是位置坐标，后 3 个参数是角度姿态。<visual name='visual'> 段定义了可视化的大小。在定义车体后，添加车子前转向轮的内容如下：

```
<collision name='caster_collision'>
  <pose>0.15 0 -0.05 0 0 0</pose>
  <geometry>
      <sphere>
      <radius>.05</radius>
    </sphere>
```

```
      </geometry>

      <surface>
        <friction>
          <ode>
            <mu>0</mu>
            <mu2>0</mu2>
            <slip1>1.0</slip1>
            <slip2>1.0</slip2>
          </ode>
        </friction>
      </surface>
    </collision>

    <visual name='caster_visual'>
      <pose>0.15 0 -0.05 0 0 0</pose>
      <geometry>
        <sphere>
          <radius>.05</radius>
        </sphere>
      </geometry>
    </visual>
  </link>
```

这里定义了一个半径为0.05的球体作为小车的前轮，轮子的位姿坐标为（0.15 0 -0.05 0 0 0）。然后添加前后轮，后轮由半径0.1长度为0.05的圆柱体组成。左右轮的位姿坐标分别为左轮（-0.1 0.18 0.1 0 1.5707 1.5707）、右轮（-0.1 -0.18 0.1 0 1.5707 1.5707）。具体内容如下：

```
    <link name="left_wheel">
      <pose>-0.1 0.18 0.1 0 1.5707 1.5707</pose>
      <collision name="collision">
        <geometry>
          <cylinder>
            <radius>.1</radius>
            <length>.05</length>
          </cylinder>
        </geometry>
      </collision>
      <visual name="visual">
        <geometry>
          <cylinder>
            <radius>.1</radius>
```

```
        <length>.05</length>
      </cylinder>
    </geometry>
  </visual>
</link>

<link name="right_wheel">
  <pose>-0.1 -0.18 0.1 0 1.5707 1.5707</pose>
  <collision name="collision">
    <geometry>
      <cylinder>
        <radius>.1</radius>
        <length>.05</length>
      </cylinder>
    </geometry>
  </collision>
  <visual name="visual">
    <geometry>
      <cylinder>
        <radius>.1</radius>
        <length>.05</length>
      </cylinder>
    </geometry>
  </visual>
</link>
```

定义左右轮关节的旋转类型为 type="revolute"，定义该关节的父子链接关系及旋转轴，内容如下：

```
<joint type="revolute" name="left_wheel_hinge">
  <pose>0 0 -0.03 0 0 0</pose>
  <child>left_wheel</child>
  <parent>chassis</parent>
  <axis>
    <xyz>0 1 0</xyz>
  </axis>
</joint>

<joint type="revolute" name="right_wheel_hinge">
  <pose>0 0 0.03 0 0 0</pose>
  <child>right_wheel</child>
  <parent>chassis</parent>
  <axis>
```

```
        <xyz>0 1 0</xyz>
      </axis>
</joint>
```

至此，车体基本机构已经构造完成。下面添加 kinect 视觉传感器，并添加长方体结构作为 kinect 与车体的连接件。添加连接件的代码如下：

```
<link name="car_kienct">
  <pose>0.15 0 0.45 0 0 0</pose>
  <collision name="collision">
    <geometry>
        <box>
          <size>0.1 0.1 0.6</size>
        </box>
    </geometry>
  </collision>
  <visual name="visual">
    <geometry>
        <box>
          <size>.1 .1 .6</size>
        </box>
    </geometry>
  </visual>
</link>
```

为连接件添加 joint 描述，旋转类型定义为 type="fixed" 不可旋转，内容如下：

```
<joint type="fixed" name="kinect_joint">
    <child>kinect::link</child>
    <parent>chassis</parent>
</joint>
```

添加 kinect 视觉传感器及其 joint 类型约束，具体内容如下：

```
        <include>
            <uri>model://kinect</uri>
            <pose>.18 0 0.8 0 0 0</pose>
        </include>

        <joint type="fixed" name="carkiect_joint">
            <child>car_kienct</child>
            <parent>chassis</parent>
        </joint>
  </model>
</sdf>
```

注意

Gazebo中的kinect模型无法与ROS进行通信。若要使其能发布ROS消息，需要修改~/.gazebo/models/kinect下的model.sdf文件内容，在</camera>下添加如下内容：

```
<plugin name="camera_plugin" filename="libgazebo_ros_openni_kinect.so">
  <baseline>0.2</baseline>
  <alwaysOn>true</alwaysOn>
  <!-- Keep this zero, update_rate in the parent <sensor> tag
    will control the frame rate. -->
  <updateRate>0.01</updateRate>
  <cameraName>camera_ir</cameraName>
  <imageTopicName>/camera/depth/image_raw</imageTopicName>
  <cameraInfoTopicName>/camera/depth/camera_info</cameraInfoTopicName>
  <depthImageTopicName>/camera/depth/image_raw</depthImageTopicName>
  <depthImageInfoTopicName>/camera/depth/camera_info</depthImageInfoTopicName>
  <pointCloudTopicName>/camera/depth/points</pointCloudTopicName>
  <frameName>camera_link</frameName>
  <pointCloudCutoff>0.05</pointCloudCutoff>
  <distortionK1>0</distortionK1>
  <distortionK2>0</distortionK2>
  <distortionK3>0</distortionK3>
  <distortionT1>0</distortionT1>
  <distortionT2>0</distortionT2>
  <CxPrime>0</CxPrime>
  <Cx>0</Cx>
  <Cy>0</Cy>
  <focalLength>0</focalLength>
  <hackBaseline>0</hackBaseline>
</plugin>
```

至此，车体搭建及kinect视觉传感器搭载完成。运行roscore，并启动Gazebo：

```
$ roscore
$ roslaunch gazebo_ros empty_world.launch
```

利用insert->My Robot即可添加机器人模型至中间可视化区域，如图3-18所示。

为了模拟真实环境，通过insert添加Brick Box 3×3×1模型墙体，通过Book shelf添加书柜，创建的仿真环境如图3-19所示。

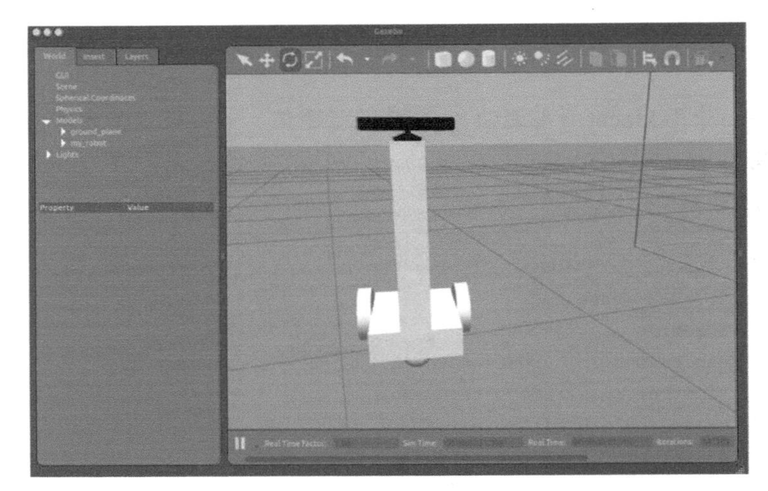

图 3-18 机器人模型

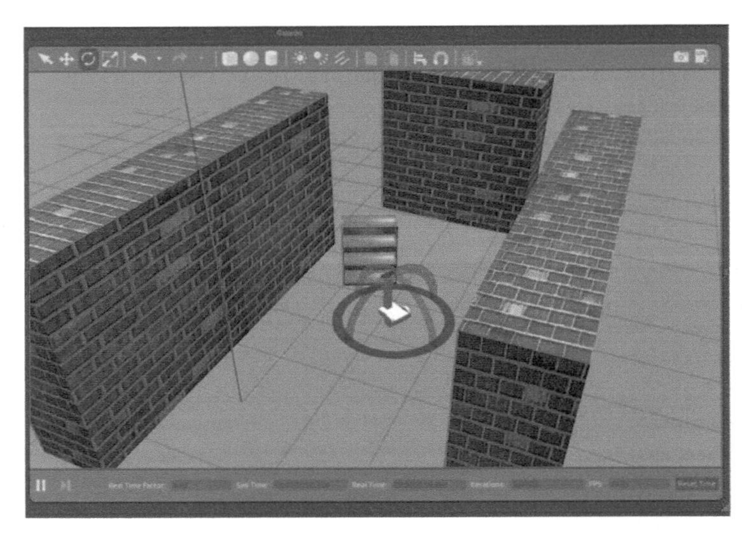

图 3-19 机器人仿真环境

查看所有的 topic 信息，运行 rostopic list：

```
/camera/depth/camera_info
/camera/depth/image_raw
/camera/depth/image_raw/compressed
/camera/depth/image_raw/compressed/parameter_descriptions
/camera/depth/image_raw/compressed/parameter_updates
/camera/depth/image_raw/compressedDepth
```

```
/camera/depth/image_raw/compressedDepth/parameter_descriptions
/camera/depth/image_raw/compressedDepth/parameter_updates
/camera/depth/image_raw/theora
/camera/depth/image_raw/theora/parameter_descriptions
/camera/depth/image_raw/theora/parameter_updates
/camera/depth/points
/camera_ir/depth/camera_info
/camera_ir/parameter_descriptions
/camera_ir/parameter_updates
/clicked_point
/clock
/gazebo/link_states
/gazebo/model_states
/gazebo/parameter_descriptions
/gazebo/parameter_updates
/gazebo/set_link_state
/gazebo/set_model_state
/initialpose
/move_base_simple/goal
/rosout
/rosout_agg
/tf
```

运行 Rviz 来查看发布的点云和图像信息，如图 3-20 所示。

图 3-20　点云和图像信息

运行 rosrun rqt_graph rqt_gtaph，查看节点关系图，如图 3-21 所示。

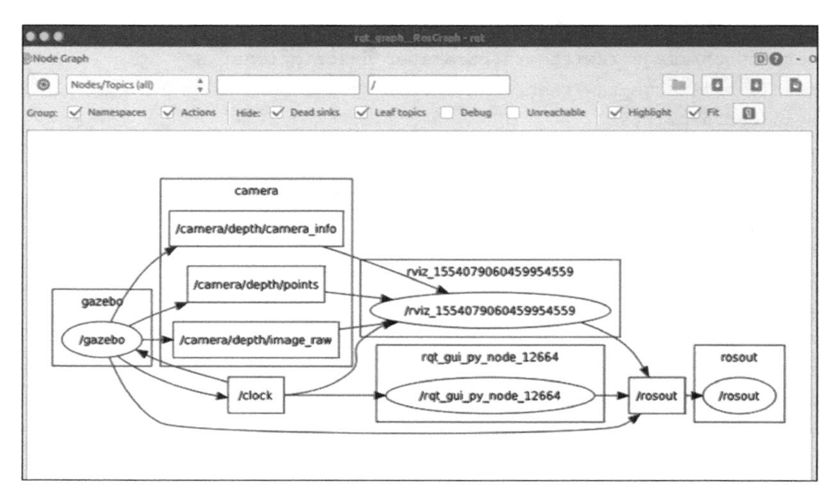

图 3-21　节点关系图

通过以上介绍，相信读者对 Gazebo 的基本使用与功能已经有了大致的了解，能够创建简单的机器人模型文件，并进行 ROS 通信，也掌握了 ROS 和 Gazebo 仿真调试的基础知识。现在，我们对使用 Gazebo 进行机器人模拟及 ROS 通信的步骤进行总结。

（1）加载 world 文件。

（2）记载机器人模型文件。

（3）设置机器人在物理世界中的位置。

（4）发布机器人本体及具体关节状态信息。

（5）打开 Rviz，观察调试机器人。

3.3 rqt的调试

rqt 是一个基于 Qt 的 ROS GUI 开发框架，以插件的形式实现各种 GUI 工具。可以将所有现有的 GUI 工具作为 rqt 中的可停靠窗口运行。使用 rqt 工具使得用户更容易地在屏幕上管理所有的窗口。rqt 由以下 3 个部分组成。

- rqt ： 核心模块。
- rqt_common_plugins：ROS 后端工具套件，可用于机器人运行时的开/关。

- rqt_robot_plugins：在运行时与机器人交互的工具。

安装 rqt 可通过以下命令完成：

```
$ sudo apt-get install ros-<distro>-rqt
$ sudo apt-get install ros-<distro>-rqt-common-plugins
```

其中，<distro> 更换为版本号，如 indigo、kinetic 等。可以单独运行 rqt 中的某一插件，如之前在查看节点关系图所使用的 rosrun rqt_graph rqt_graph。

下面以 ROS 中自带的 turtlesim 海龟功能包为例，探索 rqt 插件的具体用法。首先，运行 roscore，启动 turtlesim(rosrun turtlesim turtlesim_node)，可以看到弹出的小海龟窗口，如图 3-22 所示。

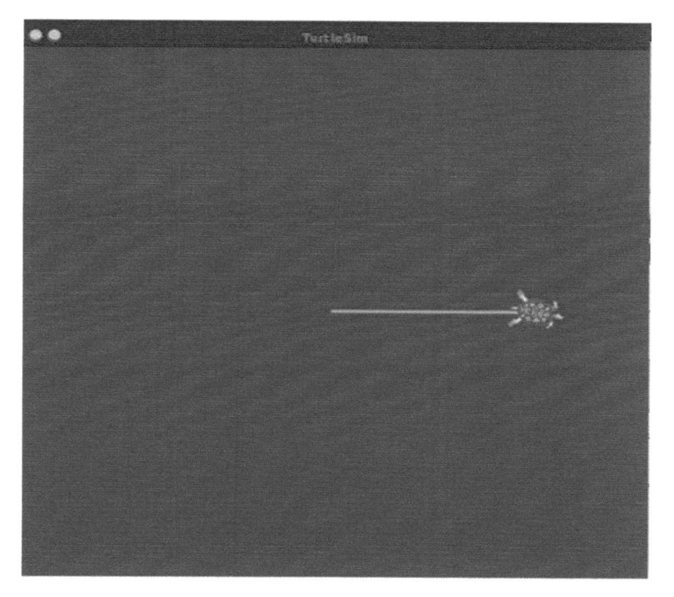

图 3-22　turtlesim 海龟

然后，使用 rqt_console 连接 ROS 的日志框架，来显示节点的输出信息。使用 rqt_logger_level 改变运行节点的等级（DEBUG、WARN、INFO、ERROR）。打开两个终端，分别运行：

```
$ rosrun rqt_console rqt_console
```

可以看到弹出的 rqt_console 窗口，如图 3-23 所示。

图 3-23　rqt_console 窗口

```
$ rosrun rqt_logger_level rqt_logger_level
```

可以看到图 3-24 所示的 rqt_logger_level 窗口。通过 rqt_logger_level 可以查看并改变运行节点的等级。

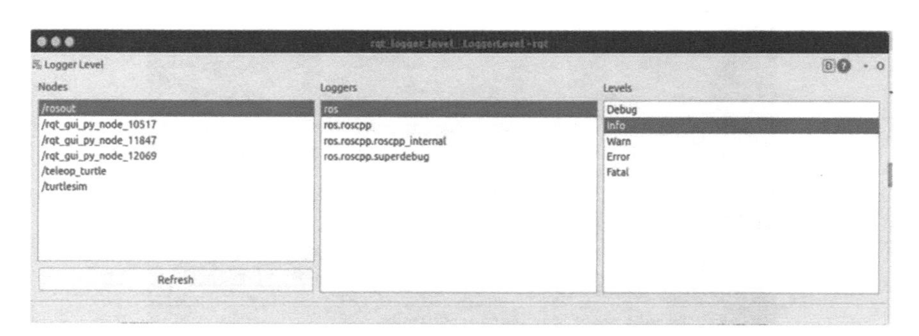

图 3-24　rqt_logger_level 窗口

发布海龟运动的线速度和角速度，控制海龟做圆周运动，在终端通过命令行发布速度 topic：

```
rostopic pub /turtle1/cmd_vel geometry_ms/Twist -r 1 -- '{linear: {x: 1.0, y: 1.0, z: 0.0}, angular: {x: 0.0,y: 0.0,z: 1.0}}'
```

可以看到做圆周运动的小海龟，如图 3-25 所示。

通过 rqt_plot 查看 topic 的变化情况：

```
$ rosrun rqt_plot rqt_plot
```

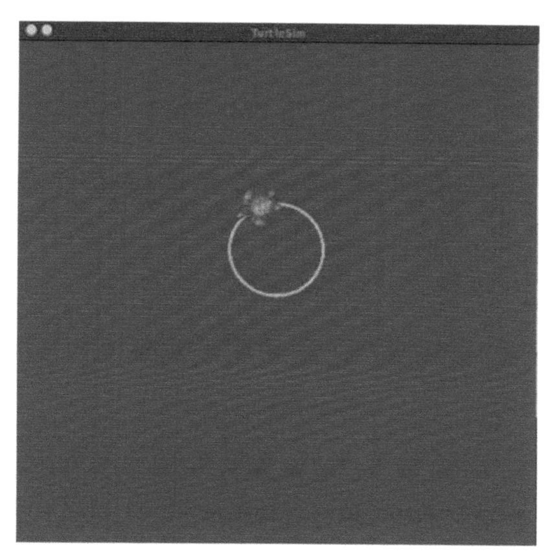

图 3-25　海龟做圆周运动

在弹出的窗口的左上角 Topic 文本框中输入 /turtle1/pose，并单击右边的“+”按钮，可以看到小海龟的线速度、角速度、x 和 y 方向的坐标值的变化情况，如图 3-26 所示。

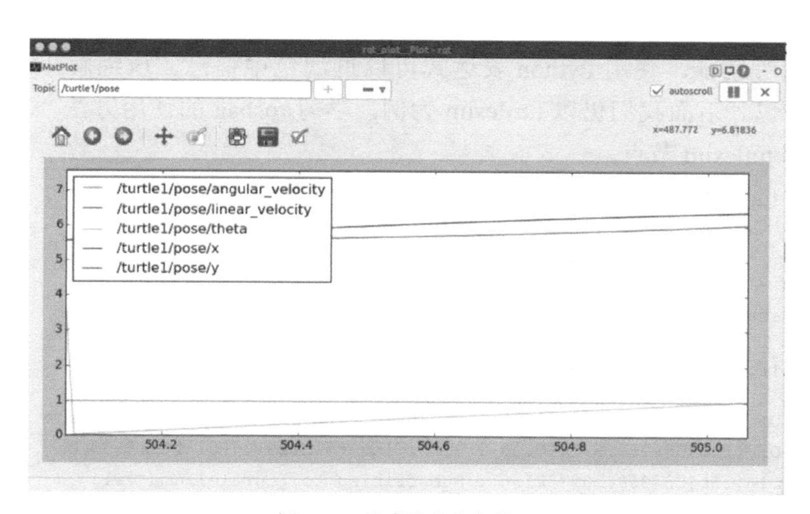

图 3-26　速度位移变化情况

运行 rqt_graph，查看节点关系图，如图 3-27 所示。

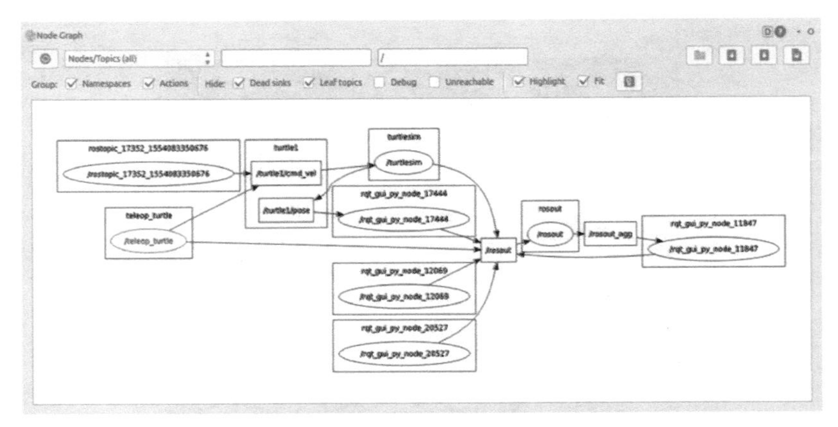

图 3-27 节点关系图

通过以上的介绍，相信读者已经掌握了 rqt 调试的常用基本功能。接下来，我们将介绍 ROS 中另一种调试工具——rosbag。

3.4 rosbag的使用

rosbag package 提供了一个命令行工具，以及 cpp 类和 Python 的 API。rosbag 使用命令行能够实现录制、从包重新发布、获取包的概括信息，以及检查包的消息类型等功能。使用 Python 表达式可以过滤包中信息、压缩解压缩包，并重新索引包。下面我们仍以 turtlesim 为例，学习 rosbag 的使用方法。首先启动 roscore 和 tutlesim 节点。

新建 turtlesim_bag 目录用于保存信息。

```
$ mkdir ~/turtlesim_bag
$ cd ~/turtlesim_bag
$ rosbag record -a
```

可以看到输出的具体信息：

```
[ INFO] [1554084960.687640036]: Subscribing to /turtle1/color_sensor
[ INFO] [1554084960.691418109]: Recording to 2019-04-01-10-16-00.bag.
[ INFO] [1554084960.709732166]: Subscribing to /turtle1/cmd_vel
[ INFO] [1554084960.720902541]: Subscribing to /rosout
[ INFO] [1554084960.730910531]: Subscribing to /rosout_agg
[ INFO] [1554084960.745193151]: Subscribing to /turtle1/pose
```

读者可以自定义 record 记录时间，若记录数据满足时间要求，可通过在运行 rosbag record 命令的窗口中按组合键 <Ctrl+C> 退出该命令，结束录制。通过 ls 命令查看当前文件夹下的文件，可以看到刚刚录制的消息记录包。

```
2019-04-01-10-16-00.bag
```

然后我们使用 rosbag info 命令来查看内容，使用 rosbag play 命令将内容回放出来。

```
$ rosbag info 2019-04-01-10-16-00.bag
```

可以看到如下输出信息：

```
path:        2019-04-01-10-16-00.bag
version:     2.0
duration:    29.1s
start:       Apr 01 2019 10:16:00.77 (1554084960.77)
end:         Apr 01 2019 10:16:29.89 (1554084989.89)
size:        261.1 KB
messages:    3637
compression: none [1/1 chunks]
types:       geometry_msgs/Twist  [9f195f881246fdfa2798d1d3eebca84a]
             rosgraph_msgs/Log    [acffd30cd6b6de30f120938c17c593fb]
             turtlesim/Color      [353891e354491c51aabe32df673fb446]
             turtlesim/Pose       [863b248d5016ca62ea2e895ae5265cf9]
topics:      /rosout                  4 msgs    : rosgraph_msgs/Log
(2 connections)
             /turtle1/cmd_vel        29 msgs    : geometry_msgs/Twist
             /turtle1/color_sensor 1803 msgs    : turtlesim/Color
             /turtle1/pose         1801 msgs    : turtlesim/Pose
```

终端运行如下：

```
$ rosbag play 2019-04-01-10-16-00.bag
```

可以看到内容回放如下：

```
[ INFO] [1554085316.844978630]: Opening 2019-04-01-10-16-00.bag
Waiting 0.2 seconds after advertising topics... done.
Hit space to toggle paused, or 's' to step.
[RUNNING]  Bag Time: 1554084981.144269   Duration: 20.378871 / 29.127471
```

在默认模式下，rosbag play 命令在公告每条消息后会等待一小段时间（0.2s），然后才真正开始发布 bag 文件中的内容。等待的过程可以通知消息订阅器，消息已经公告了，消息数据可能会马上到来。

rosbag record 命令支持只录制某些特别指定的话题到单个 bag 文件中，这样就允许用户录制特定的感兴趣的话题。如果读者只想录制 turtlesim 的速度和位置话题，即 /turtle1/cmd_vel 和 /turtle1/pose，那么可以通过如下命令完成：

```
rosbag record -O bag_velandpose /turtle1/cmd_vel /turtle1/pose
```

上述命令中的 -O 参数告诉 rosbag record 将数据记录保存到名为 bag_velandpose 的文件中，同时后面的话题参数用于告诉 rosbag record 只能录制 /turtle1/cmd_vel 和 /turtle1/pose 这两个指定的话题。通过 ls 命令查看当前目录下的 bag 文件，可以看到新录制的文件如下：

```
2019-04-01-10-16-00.bag  bag_velandpose.bag
```

3.5 rosbridge的开发

rosbridge（rosbridge_suite）是 ROS 官方为开发者提供的一个用于非 ROS 系统和 ROS 系统之间交互通信的功能包。rosbridge 主要包含两个部分：rosbridge Protocol 和 rosbridge Implementation。其中，Protocol 部分提供了非 ROS 系统和 ROS 系统通信的具体格式，包括话题的订阅、消息的发布、服务的调用、参数的设置和获取、图片信息的传递等，都是 JSON 格式的字符串；Implementation 部分是 rosbridge 的具体实现，包含 rosapi、rosbridge_library、rosbridge_server 等包。rosapi 通过服务调用使某些 ROS action 可访问，包括获取和设置参数、获取主题列表等。rosbridge_library 是核心 rosbridge 包。rosbridge_library 负责获取 JSON 字符串并将命令发送到 ROS，也可以将 ROS 消息话题等转换成 JSON 字符串。rosbridge_server 负责通信的传输层，包括 websocket、tcp、udp 等几种格式。下面我们以测试 ROS 与浏览器网页的通信为例，简单介绍 rosbridge 的使用。

3.5.1 rosbridge_suite的安装

使用 rosbridge 前，我们需要先安装 rosbridge_suite。rosbridge_suite 是一个包集合，主要包含以下 3 个子功能包。

- rosbridge_library：rosbridge 核心包。rosbridge_library 负责接收 JSON 命令并将指令发送给 ROS，同样，由 ROS 发送的指令会由其转换为 JSON 字符串发出。

- rosapi：使某个 ROS 可执行程序可以访问为 ROS client libraries 准备的服务，包括获取和设置参数、获取主题列表等。
- rosbridge_server：提供 JSON <-> ROS 的转换。rosbridge_server 提供了一个 websocket 连接，使浏览器可以与 rosbridge 进行通信。

可通过如下命令安装 rosbridge-suite：

```
sudo apt-get install ros-<rosdistro>-rosbridge-suite
```

<rosdistro> 为对应的 ROS 版本，如 indigo 等。

> **注意**
>
> rosbridge-suite 已经包含了 rosbridge-server，不需要单独安装。（如果读者只安装 rosbridge-server，那么可以通过单独安装命令 sudo apt-get install ros-<rosdistro>-rosbridge-server 进行安装。）

自定义开发过程的读者可以自己下载源码，在 catkin 工作空间中进行编译安装。这样就可以自定义 Launch 文件，例如修改 WebSocket 默认端口号 9090 为 8080，修改 ~/catkin_ws/src/rosbridge_suite/rosbridge_server/launch 目录下的 rosbridge_websocket.launch 文件，将 <arg name="port" default="9090" /> 的 9090 改为 8080 即可。自定义编译过程如下：

```
$ sudo apt-get install ros-indigo-rosauth
$ source /opt/ros/indigo/setup.bash
$ cd ~/catkin_ws/src
$ git clone https://github.com/RobotWebTools/rosbridge_suite.git
$ cd ../
$ catkin_make
```

3.5.2 测试html通信

创建 sample.html 文件，打开文件并添加如下内容：

```
<!DOCTYPE html>
<html>
<head>
<meta charset="utf-8" />
<script src="https://static.robotwebtools.org/EventEmitter2/current/
eventemitter2.min.js"></script>
<script type="text/javascript" src="http://static.robotwebtools.org/roslibjs/
current/roslib.min.js"></script>
```

```
    <script src="https://github.com/RobotWebTools/roslibjs/blob/develop/build/
roslib.js"></script>

    <script>
      // Connecting to ROS
      // -----------------
      var ros = new ROSLIB.Ros();
      // If there is an error on the backend, an 'error' emit will be emitted.
      ros.on('error', function(error) {
        document.getElementById('connecting').style.display = 'none';
        document.getElementById('connected').style.display = 'none';
        document.getElementById('closed').style.display = 'none';
        document.getElementById('error').style.display = 'inline';
        console.log(error);
      });
      // Find out exactly when we made a connection.
      ros.on('connection', function() {
        console.log('Connection made!');
        document.getElementById('connecting').style.display = 'none';
        document.getElementById('error').style.display = 'none';
        document.getElementById('closed').style.display = 'none';
        document.getElementById('connected').style.display = 'inline';
      });
      ros.on('close', function() {
        console.log('Connection closed.');
        document.getElementById('connecting').style.display = 'none';
        document.getElementById('connected').style.display = 'none';
        document.getElementById('closed').style.display = 'inline';
      });
      // Create a connection to the rosbridge WebSocket server.
      ros.connect('ws://localhost:9090');
      // Publishing a Topic
      // ------------------
      // First, we create a Topic object with details of the topic's name and
      // message type.
      var cmdVel = new ROSLIB.Topic({
        ros : ros,
        name : '/cmd_vel',
        messageType : 'geometry_msgs/Twist'
      });
      // Then we create the payload to be published. The object we pass in to ros.Message
      // matches the fields defined in the geometry_msgs/Twist.msg definition.
      var twist = new ROSLIB.Message({
        linear : {
```

```
    x : 0.1,
    y : 0.2,
    z : 0.3
  },
  angular : {
    x : -0.1,
    y : -0.2,
    z : -0.3
  }
});
// And finally, publish.
cmdVel.publish(twist);
//Subscribing to a Topic
//----------------------
// Like when publishing a topic, we first create a Topic object with details of
// the topic's name and message type. Note that we can call publish or subscribe
// on the same topic object.
var listener = new ROSLIB.Topic({
  ros : ros,
  name : '/listener',
  messageType : 'std_msgs/String'
});
// Then we add a callback to be called every time a message is published on
// this topic.
listener.subscribe(function(message) {
  console.log('Received message on ' + listener.name + ': ' + message.data);
  // If desired, we can unsubscribe from the topic as well.
  listener.unsubscribe();
});
// Calling a service
// -----------------
// First, we create a Service client with details of the service's name and
// service type.
var addTwoIntsClient = new ROSLIB.Service({
  ros : ros,
  name : '/add_two_ints',
  serviceType : 'rospy_tutorials/AddTwoInts'
});
// Then we create a Service Request. The object we pass in to ROSLIB.
// ServiceRequest matches the  fields defined in the rospy_tutorials AddTwoInts.
// srv file.
var request = new ROSLIB.ServiceRequest({
  a : 1,
  b : 2
```

```
    });
    // Finally, we call the /add_two_ints service and get back the results in the
    // callback. The result is a ROSLIB.ServiceResponse object.
    addTwoIntsClient.callService(request, function(result) {
      console.log('Result for service call on ' + addTwoIntsClient.name + ': ' +
result.sum);
    });
    // Advertising a Service
    // ---------------------
    // The Service object does double duty for both calling and advertising services
    var setBoolServer = new ROSLIB.Service({
      ros : ros,
      name : '/set_bool',
      serviceType : 'std_srvs/SetBool'
    });
    // Use the advertise() method to indicate that we want to provide this service
    setBoolServer.advertise(function(request, response) {
      console.log('Received service request on ' + setBoolServer.name + ': ' +
request.data);
      response['success'] = true;
      response['message'] = 'Set successfully';
      return true;
    });
    // Setting a param value
    // ---------------------
    ros.getParams(function(params) {
      console.log(params);
    });
    // First, we create a Param object with the name of the param.
    var maxVelX = new ROSLIB.Param({
      ros : ros,
      name : 'max_vel_y'
    });
    //Then we set the value of the param, which is sent to the ROS Parameter Server.
    maxVelX.set(0.8);
    maxVelX.get(function(value) {
      console.log('MAX VAL: ' + value);
    });
    // Getting a param value
    // ---------------------
    var favoriteColor = new ROSLIB.Param({
      ros : ros,
      name : 'favorite_color'
```

```
      });
      favoriteColor.set('red');
      favoriteColor.get(function(value) {
        console.log('My robot\'s favorite color is ' + value);
      });
    </script>
    </head>

    <body>
      <h1>Simple roslib Example</h1>
      <p>Run the following commands in the terminal then refresh this page. Check
the JavaScript
        console for the output.</p>
      <ol>
        <li><tt>roscore</tt></li>
        <li><tt>rostopic pub /listener std_msgs/String "Hello, World"</tt></li>
        <li><tt>rostopic echo /cmd_vel</tt></li>
        <li><tt>rosrun rospy_tutorials add_two_ints_server</tt></li>
        <li><tt>roslaunch rosbridge_server rosbridge_websocket.launch</tt></li>
      </ol>
      <div id="statusIndicator">
        <p id="connecting">
          Connecting to rosbridge...
        </p>
        <p id="connected" style="color:#00D600; display:none">
          Connected
        </p>
        <p id="error" style="color:#FF0000; display:none">
          Error in the backend!
        </p>
        <p id="closed" style="display:none">
          Connection closed.
        </p>
      </div>
    </body>
    </html>
```

说明如下。

导入所有需要的 JavaScript 文件，ros.on() 用于提示连接成功失败或关闭的信息，ros.connect 用于 ROS 通信连接并定义 WebSocket 在端口 9090。

- var cmdVel = new ROSLIB.Topic： 创建 Topic 对象用于发布 cmd_vel。

- ROSLIB.Message：用于发布速度消息，在其中定义线速度和角速度。
- var listener = new ROSLIB.Topic：用于接收 '/listener' 话题上 'std_msgs/String' 类型消息。
- var addTwoIntsClient = new ROSLIB.Service：创建服务客户端。
- var request = new ROSLIB.ServiceRequest：定义服务请求的内容。
- var maxVelX = new ROSLIB.Param：创建 Param 对象，用于获取和设置参数。

结尾部分的内容用于显示调试提示信息和连接状况。

现在开始测试通信。启动 roscore，分别在终端运行如下命令：

```
$ rostopic pub /listener std_msgs/String "Hello, World"
$ rostopic echo /cmd_vel
$ rosrun rospy_tutorials add_two_ints_server
$ roslaunch rosbridge_server rosbridge_websocket.launch
```

启动完 rosbridge_server 后，可看到如下提示信息：

```
SUMMARY
========

PARAMETERS
 * /rosapi/params_glob: [*]
 * /rosapi/services_glob: [*]
 * /rosapi/topics_glob: [*]
 * /rosbridge_websocket/address:
 * /rosbridge_websocket/authenticate: False
 * /rosbridge_websocket/bson_only_mode: False
 * /rosbridge_websocket/delay_between_messages: 0
 * /rosbridge_websocket/fragment_timeout: 600
 * /rosbridge_websocket/max_message_size: None
 * /rosbridge_websocket/params_glob: [*]
 * /rosbridge_websocket/port: 9090
 * /rosbridge_websocket/retry_startup_delay: 5
 * /rosbridge_websocket/services_glob: [*]
 * /rosbridge_websocket/topics_glob: [*]
 * /rosbridge_websocket/unregister_timeout: 10
 * /rosdistro: indigo
 * /rosversion: 1.11.21

NODES
  /
    rosapi (rosapi/rosapi_node)
```

```
    rosbridge_websocket (rosbridge_server/rosbridge_websocket)

ROS_MASTER_URI=http://192.168.43.187:11311
core service [/rosout] found
process[rosbridge_websocket-1]: started with pid [13333]
process[rosapi-2]: started with pid [13334]
registered capabilities (classes):
 - rosbridge_library.capabilities.call_service.CallService
 - rosbridge_library.capabilities.advertise.Advertise
 - rosbridge_library.capabilities.publish.Publish
 - rosbridge_library.capabilities.subscribe.Subscribe
 - <class 'rosbridge_library.capabilities.defragmentation.Defragment'>
 - rosbridge_library.capabilities.advertise_service.AdvertiseService
 - rosbridge_library.capabilities.service_response.ServiceResponse
 - rosbridge_library.capabilities.unadvertise_service.UnadvertiseService
the rosdep view is empty: call 'sudo rosdep init' and 'rosdep update'
[INFO] [WallTime: 1554095878.425031] Rosbridge WebSocket server started on port 9090
```

下面用谷歌 Chrome 浏览器打开 sample.html 文件，可看到 rosbridge_server 提示更新如下：

```
[INFO] [WallTime: 1554096003.028570] Client connected.  1 clients total.
the rosdep view is empty: call 'sudo rosdep init' and 'rosdep update'
[INFO] [WallTime: 1554096003.195668] [Client 0] Subscribed to /listener
[INFO] [WallTime: 1554096003.200418] [Client 0] Advertised service /set_bool.
[INFO] [WallTime: 1554096003.232310] [Client 0] Unsubscribed from /listener
```

从终端的提示信息中可以看到通信连接成功，此时浏览器界面状态如图 3-28 所示。

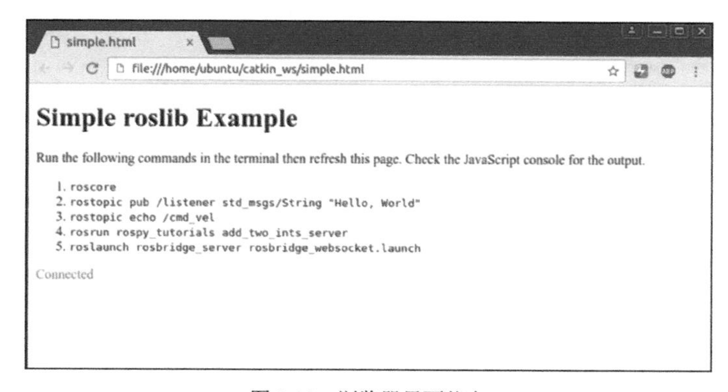

图 3-28　浏览器界面状态

运行 rostopic echo /cmd_vel 和 rosrun rospy_tutorials add_two_ints_server，终端提示信息如图 3-29 所示。

```
ubuntu@ubuntu-Lenovo-IdeaPad-Y400:~$ rostopic echo /cmd_vel
WARNING: topic [/cmd_vel] does not appear to be published yet
linear:
  x: 0.1
  y: 0.2
  z: 0.3
angular:
  x: -0.1
  y: -0.2
  z: -0.3
---

ubuntu@ubuntu-Lenovo-IdeaPad-Y400:~$ rosrun rospy_tutorials add_two_ints_server
Returning [1 + 2 = 3]
```

图 3-29 终端提示信息

在浏览器网页界面右击选择“inspect”，然后单击上方的“Console”，可以看到接收到的打印信息，如图 3-30 所示。

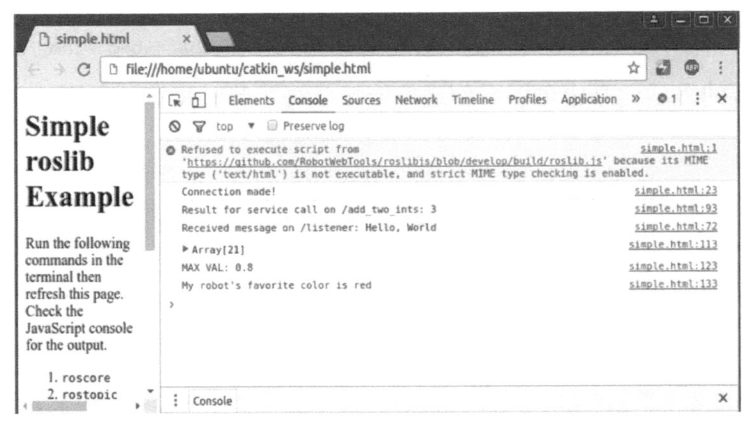

图 3-30 接收打印信息

至此，我们已经学习并练习了有关 rosbridge 通信的基础知识。若读者对这部分内容感兴趣，可查阅相关资料，做进一步的开发与研究。

第 4 章

TF 简介及应用

4.1 TF包概述

TF（transform）是一个允许用户时刻跟踪多个坐标框架的包，能及时地维持树结构中坐标帧之间的关系，并允许用户在任何时间点变换任何两个坐标帧之间的点和向量等。自 ROS Hydro 出现以来，TF 已被“弃用”，转而支持 TF2。TF2 是 TF 的新版，在更加有效地提供原有功能的基础上，添加了一些新功能。在 TF2 中，TF 中原有的 API 保持其原来的形式。由于 TF2 有 TF 特征的超集及其依赖关系的子集，因此，TF 所实现的功能在底层上已经由调用 TF2 中的函数所代替，这意味着所有用户都能够与 TF2 兼容。建议初学者直接使用 TF2，因为它有更加简洁好用的接口。机器人系统通常具有许多随时间变化的三维坐标系，例如世界坐标系、基坐标系、机械臂坐标系和头部坐标系等。TF 时刻保持记录所有这些坐标系之间的关系。图 4-1 为 PR2 机器人的 TF 坐标系。

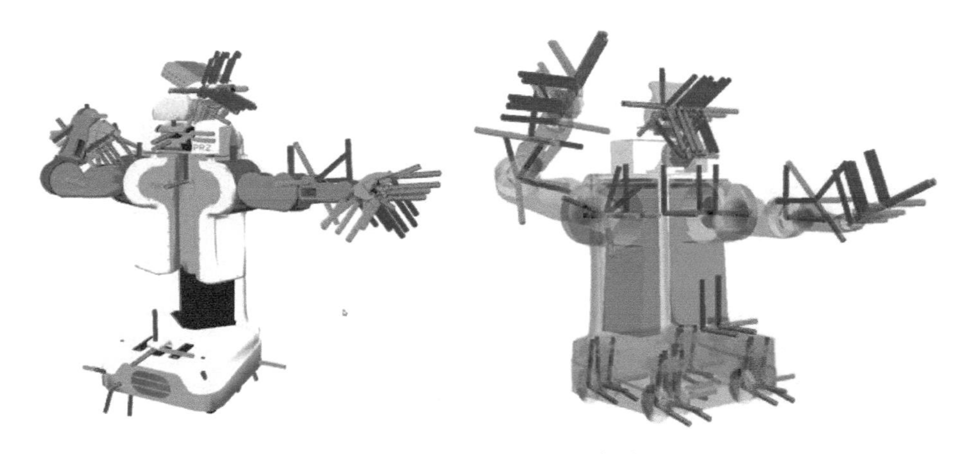

图 4-1　PR2 机器人的 TF 坐标系

4.2 TF包的简单使用

本节我们将简单使用 ROS 官网提供的功能包，并体验 TF 功能包的作用。首先，安装功能包，安装命令如下：

```
$ sudo apt-get install ros-indigo-ros-tutorials ros-indigo-geometry-
tutorials ros-indigo-rviz ros-indigo-rosbash ros-indigo-rqt-tf-tree
```

运行 turtle 功能包：

```
roslaunch turtle_tf turtle_tf_demo.launch
```

可以看到弹出窗口中有两个重叠的海龟，如图 4-2 所示。

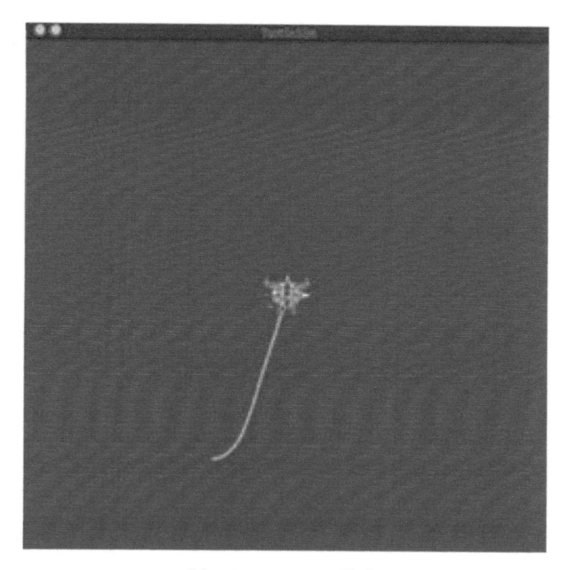

图 4-2　turtlesim 海龟

此时，注意 roslaunch turtle_tf turtle_tf_demo.launch 终端中的消息如下：

```
core service [/rosout] found
process[sim-1]: started with pid [19300]
process[teleop-2]: started with pid [19301]
process[turtle1_tf_broadcaster-3]: started with pid [19302]
process[turtle2_tf_broadcaster-4]: started with pid [19304]
Reading from keyboard
---------------------------
Use arrow keys to move the turtle.
```

言外之意，我们可以通过键盘上的上下左右按键控制小海龟移动。移动其中一只海龟，你将发现另一只小海龟也跟随移动，轨迹如图 4-3 所示。

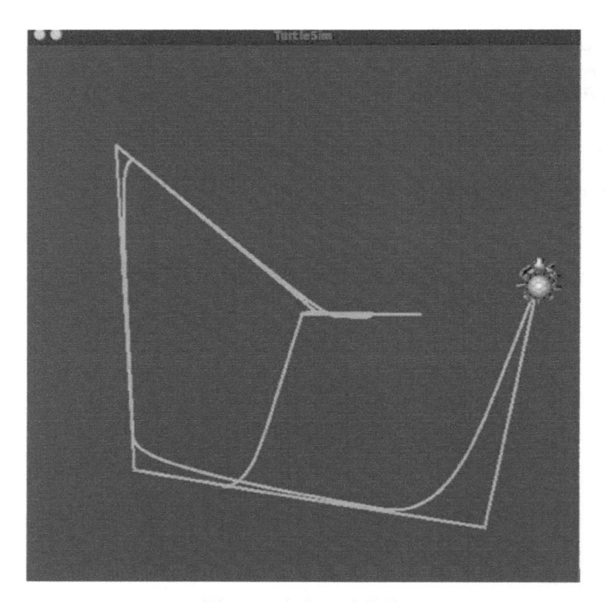

图 4-3 海龟运动轨迹

下面我们利用 ROS 提供的 tf tools 来监视 TF 信息。利用 view_frames，运行如下代码：

```
$ rosrun tf view_frames
the rosdep view is empty: call 'sudo rosdep init' and 'rosdep update'
Listening to /tf for 5.000000 seconds
Done Listeningtransfpub.sendTransform(tf::StampedTransform(base2laser,ros::Time::
now(),"base_link","laser_link"));
dot - graphviz version 2.36.0 (20140111.2315)

Detected dot version 2.36
frames.pdf generated
```

通过命令 $ evince frames.pdf 来查看生成的 pdf（读者也可以任意选择），内容如图 4-4 所示。

除了 tf view_frames 工具，我们也可以使用 rqt_tf_tree 来查看 TF 树，命令如下：

```
rosrun rqt_tf_tree rqt_tf_tree (rqt &)
```

结果如图 4-5 所示。

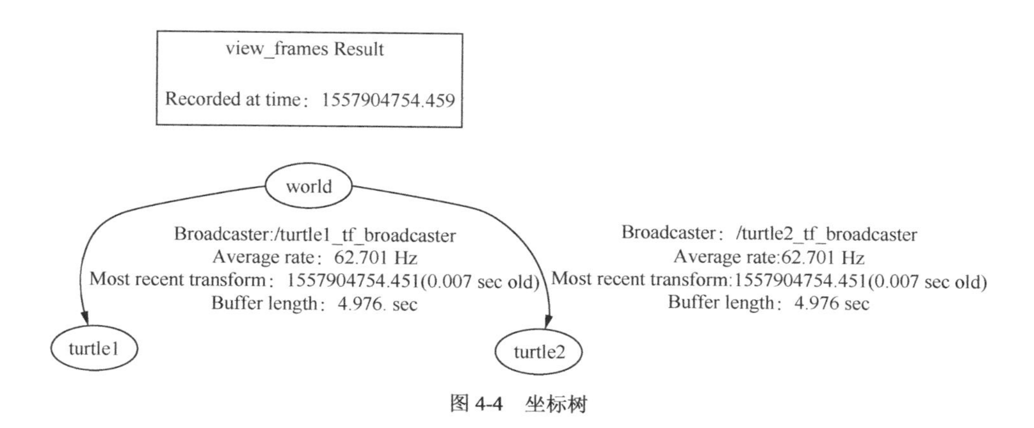

图 4-4 坐标树

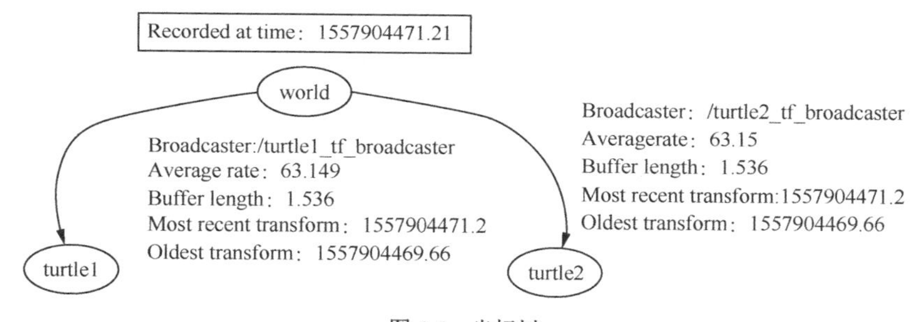

图 4-5 坐标树

仔细观察图 4-4 与图 4-5，我们发现两个框架（frame）之间有一个 broadcaster，为了使两个 frame 之间能够正确连通，中间会有一个节点来发布消息给 broadcaster。如果缺少节点来发布消息维护连通，那么这两个框架之间的连接就会断掉。broadcaster 就是一个发布者，如果两个框架之间发生了相对运动，broadcaster 就会发布相关消息。

接下来，我们利用 ROS 中的另一个工具 tf_echo 来监视两个坐标系之间的坐标变换关系，使用方法如下：

```
rosrun tf tf_echo [reference_frame] [target_frame]
```

根据上述用法查看 turtle2 坐标系相对于 turtle1 坐标系的关系，其几何关系如下：

$$T_{turtle1\_turtle2} = T_{turtle1\_world} * T_{world\_turtle2}$$

查看命令如下：

```
$ rosrun tf tf_echo turtle1 turtle2
```

此时，在终端中可以看到如下打印消息：

```
At time 1557906316.466
- Translation: [0.000, 0.000, 0.000]
- Rotation: in Quaternion [0.000, 0.000, 0.061, 0.998]
            in RPY (radian) [0.000, -0.000, 0.122]
            in RPY (degree) [0.000, -0.000, 7.010]
At time 1557906317.234
- Translation: [0.000, 0.000, 0.000]
- Rotation: in Quaternion [0.000, 0.000, 0.061, 0.998]
            in RPY (radian) [0.000, -0.000, 0.122]
            in RPY (degree) [0.000, -0.000, 7.010]
```

读者可利用键盘控制海龟移动，这时打印出的坐标变换关系也会实时地发生变化。最后，我们利用 ROS 中常用的调试根据 Rviz 来查看 TF 树。

```
rosrun rviz rviz
```

在可视化界面中选择 TF 并添加（如果坐标系太小，适当增大 marker scale 的大小以便查看），可以看到坐标系关系如图 4-6 所示。

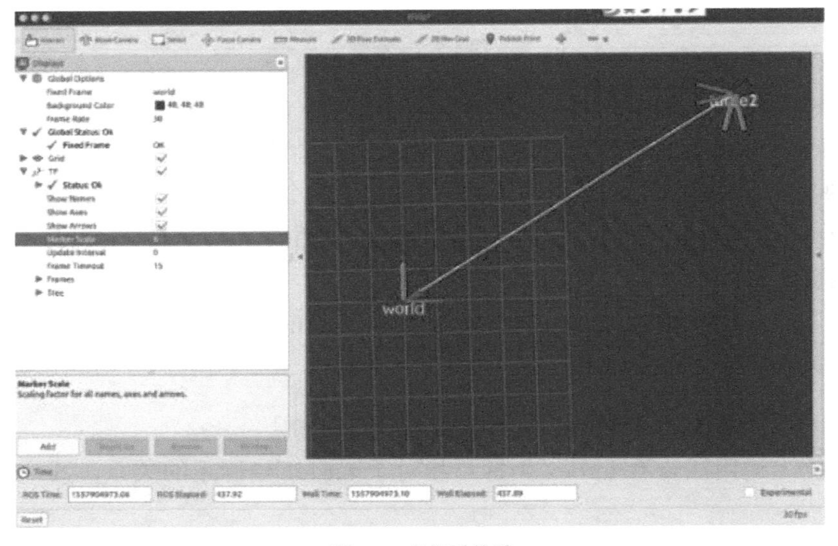

图 4-6　坐标系关系

4.3 编写TF发布与接收程序

在使用了 ROS 中自带的功能包之后，接下来我们需要自己编写 TF 的发布和接收程序，只有这样才能更深刻理解并掌握 TF 的功能。在编写程序之前，我们需要了解 TF 的消息格式——TransformStampde.msg，它是处理两个框架之间 TF 关系的数据格式。具体内容如下：

```
std_mags/Header header
        uint32 seq
        time stamp
        string frame_id
string child_frame_id
geometry_msgs/Transform transform
        geometry_msgs/Vector3 translation
                float64 x
                float64 y
                float64 z
        geometry_msgs/Quaternion rotation
                float64 x
                float64 y
                flaot64 z
                float64 w
```

在上述消息格式中，header 定义了序号、时间以及父框架的名称。String 定义了 child_frame，父坐标系和自坐标系之间变换关系由 geometry_msgs/Transform 来定义。在 transform 下面又分别通过 geometry_msgs/Vector3 和 geometry_msgs/Quaternion 定义了两个坐标系之间的平移和旋转尺度。

ROS 的 TF 中定义了点、向量、四元数、坐标变换、矩阵、位姿等数据格式。下面列举了常用的数据格式：

```
tf::Point, tf::Vector3, tf::Quaternion, tf::Matrix3x3, tf::pose, tf::Transform
```

若想对数据添加时间戳，只需在上述数据前加上 Stamped，例如，tf:Stamped Transform 即为带时间戳的坐标变换。

发布 TF 变换需要 tf::TransformBroadcaster 类，实际使用时调用该类中 sendTransform 函数即可。该函数的重载有以下几种形式，可根据实际需要进行选择。

```
void sendTransform(const StampedTransform &transform)
void sendTransform(const std::vector<StampedTransform> &transforms)
void sendTransform(const geometry_msgs::TransformStamped &transform)
void sendTransform(const std::vector<geometry_msgs::TransformStamped> &transforms)
```

有了发布者，自然不能缺少接收者，接收 TF 变换主要使用 tf::TransformListener 类。该类中常用的方法如下：

```
void    lookupTransform (const std::string &target_frame, const std::string
&source_frame, const ros::Time &time, StampedTransform &transform) const
Get the transform between two frames by frame ID.
bool    canTransform (const std::string &target_frame, const std::string
&source_frame, const ros::Time &time, std::string *error_msg=NULL) const
    Test if a transform is possible.
bool    waitForTransform (const std::string &target_frame, const std::string
&source_frame, const ros::Time &time, const ros::Duration &timeout, const ros::Duration
&polling_sleep_duration=ros::Duration(0.01), std::string *error_msg=NULL) const
    Block until a transform is possible or it times out.
```

上述 3 种方法的主要作用分别为：获取两个坐标系之间的变换、测试两个坐标系是否可以进行坐标变换、等待两个坐标系之间的变换关系连接。

在进行以上内容的准备和铺垫之后，现在我们可以编写程序了。来看一个简单的例子，将一个坐标系中的点转换到另一个坐标系中。

首先，编写 TF 发布程序。创建 book_tfpub.cpp 文件，内容如下：

```
#include "ros/ros.h"
#include "tf/transform_broadcaster.h"
#include "geometry_msgs/Point.h"
#include "tf/tf.h"

int main(int argc,char** argv)
{
    //initialize node and name
    ros::init(argc,argv,"tf_transformpub");
    //create node handle
    ros::NodeHandle nh;
    //create transform broadcast
    static tf::TransformBroadcaster transfpub;
    //create transform
    tf::Transform base2laser;
    //set translation
    base2laser.setOrigin(tf::Vector3(1,0,0));
```

```
    //create transform quaternion 0 degree
    tf::Quaternion q;
    q.setRPY(0,0,0);
    //set transform
    base2laser.setRotation(q);
    ros::Rate rate(10);

    while(nh.ok())
    {
        //loop pub transform
         transfpub.sendTransform(tf::StampedTransform(base2laser,ros::Time::now
(),"base_link","laser_link"));
        rate.sleep();
    }

    return 0;
}
```

说明如下。

- static tf::TransformBroadcaster transfpub：实例化 TransformBroadcaster 类对象。
- tf::Transform base2laser：定义基坐标系到雷达坐标系的变换关系。
- base2laser.setOrigin(tf::Vector3(1,0,0))：定义 laser 坐标系原点在 base 坐标系的（1,0,0）点，即 x 轴前方 1m。
- transfpub.sendTransform(tf::StampedTransform(base2laser,ros::Time::now(), "base_link","laser_link"))：在循环体中以 10Hz 的频率发布坐标变换关系。

然后新建 book_tflis.cpp 文件用于 TF 接收者程序，内容如下：

```
#include "ros/ros.h"
#include "tf/transform_listener.h"
#include "geometry_msgs/PointStamped.h"

using namespace std;

int main(int argc,char** argv)
{
    //initial and name node
    ros::init(argc,argv,"tf_transformlis");
    //create nodehandle
    ros::NodeHandle nh;
    //create tf listener
```

```
    tf::TransformListener tflis;
    //create point in laser_link
    geometry_msgs::PointStamped plaser;
    plaser.header.frame_id="laser_link";
    plaser.point.x=1,plaser.point.y=0,plaser.point.z=0;
    //create point in base_link
    geometry_msgs::PointStamped pbase;
    //create frequency
    ros::Rate r(10);
    while(nh.ok())
    {
        cout<<"start listening"<<endl;
        tflis.waitForTransform("base_link","laser_link",ros::Time(0),ros::Duration(3));
        tflis.transformPoint("base_link",plaser,pbase);
        cout<<"pbase is:"<<"("
            <<pbase.point.x<<","
            <<pbase.point.y<<","
            <<pbase.point.z<<")"<<endl;
        r.sleep();

    }
    return 0;
}
```

说明如下。

- tf::TransformListener tflis：实例化 TransformListener 类对象。
- geometry_msgs::PointStamped plaser：定义 laser_link 坐标系中的点。
- geometry_msgs::PointStamped pbase：定义 base_link 坐标系中的点。
- tflis.waitForTransform("base_link","laser_link",ros::Time(0),ros::Duration(3))：监听并获取两个坐标系之间的变换关系。
- tflis.transformPoint("base_link",plaser,pbase)：将 laser_link 坐标系中的点变换到 base_link 坐标系中。

创建 CMakeLists.txt 文件，文件内容如下：

```
cmake_minimum_required(VERSION 2.8.3)
project(book_tfpub)

find_package(catkin REQUIRED COMPONENTS roscpp rospy std_msgs tf geometry_msgs)

catkin_package(
#  INCLUDE_DIRS include
```

```
# LIBRARIES book_tfpub
# CATKIN_DEPENDS other_catkin_pkg
# DEPENDS system_lib
)

include_directories(
  include ${catkin_INCLUDE_DIRS}
# include
# ${catkin_INCLUDE_DIRS}
)

add_executable(book_tfpub
  src/book_tfpub.cpp
)
add_dependencies(book_tfpub ${${PROJECT_NAME}_EXPORTED_TARGETS} ${catkin_
EXPORTED_TARGETS})
target_link_libraries(book_tfpub
  ${catkin_LIBRARIES}
)

add_executable(book_tflis
  src/book_tflis.cpp
)
add_dependencies(book_tflis ${${PROJECT_NAME}_EXPORTED_TARGETS} ${catkin_
EXPORTED_TARGETS})
target_link_libraries(book_tflis
  ${catkin_LIBRARIES}
)
```

编译之后，在不同终端运行以下命令：

```
rosrun book_tfpub book_tfpub
rosrun book_tfpub book_tflis
```

在 listener 终端中，可以看到如下的打印消息：

```
start listening
pbase is:(2,0,0)
start listening
pbase is:(2,0,0)
start listening
pbase is:(2,0,0)
```

利用 tf_echo 查看坐标变换关系：

```
rosrun tf tf_echo base_link laser_link
```

打印内容如下：

```
At time 1557909984.444
- Translation: [1.000, 0.000, 0.000]
- Rotation: in Quaternion [0.000, 0.000, 0.000, 1.000]
            in RPY (radian) [0.000, -0.000, 0.000]
            in RPY (degree) [0.000, -0.000, 0.000]
At time 1557909985.044
- Translation: [1.000, 0.000, 0.000]
- Rotation: in Quaternion [0.000, 0.000, 0.000, 1.000]
            in RPY (radian) [0.000, -0.000, 0.000]
            in RPY (degree) [0.000, -0.000, 0.000]
At time 1557909986.044
- Translation: [1.000, 0.000, 0.000]
- Rotation: in Quaternion [0.000, 0.000, 0.000, 1.000]
            in RPY (radian) [0.000, -0.000, 0.000]
            in RPY (degree) [0.000, -0.000, 0.000]
```

利用 rqt_tf_tree 查看运行：

```
rosrun rqt_tf_tree rqt_tf_tree
```

结果如图 4-7 所示。

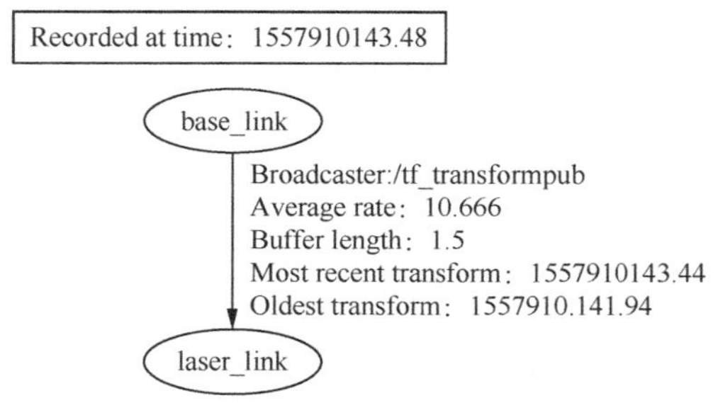

图 4-7　坐标树

运行 Rviz 查看 TF 坐标系，如图 4-8 所示。

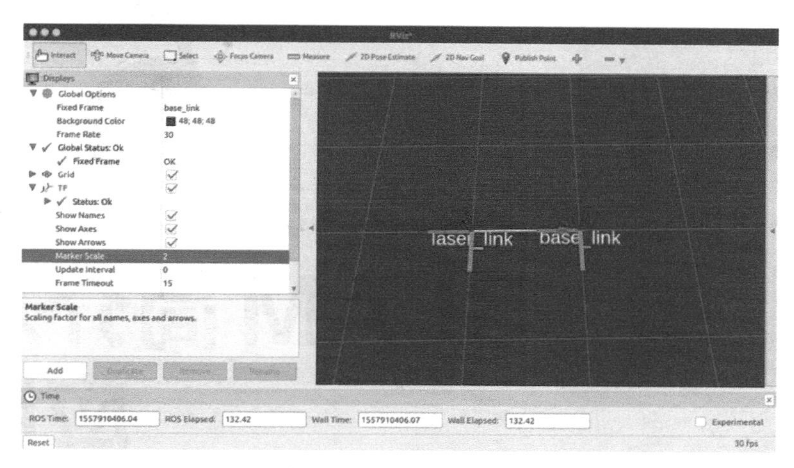

图 4-8　坐标系相对关系

查看节点关系图，运行：

```
rosrun rqt_graph rqt_graph
```

结果如图 4-9 所示。

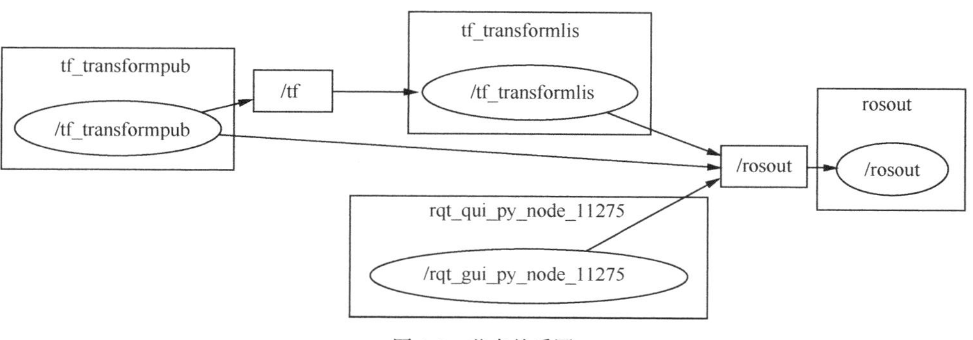

图 4-9　节点关系图

第5章

SLAM简介及应用

5.1 SLAM概述

SLAM（Simultaneous Localization and Mapping）也称为同步定位与地图构建。机器人从未知环境的未知地点出发，在运动过程中通过重复观测到的地图特征定位自身位置和姿态，再根据自身位置增量式地构建地图，从而达到同时定位和地图构建的目的。

自从20世纪80年代SLAM概念提出到现在，SLAM技术已经走过30多年的历史。SLAM系统使用的传感器也在不断升级，从早期的声呐，到后来的2D/3D激光雷达，再到单目、双目、RGBD、ToF等各种相机，以及与惯性测量单元IMU等传感器的融合。SLAM的算法也从最初的基于滤波器的方法（EKF、PF等）向基于优化的方法转变，技术框架也从开始的单一线程向多线程演进。本书主要介绍激光SLAM，视觉SLAM在此不作讨论。

基于激光雷达的SLAM（Lidar SLAM）采用2D或3D激光雷达（也叫单线或多线激光雷达）。室内机器人（如扫地机器人）一般使用2D激光雷达；无人驾驶领域一般使用3D激光雷达。各种激光雷达如图5-1所示。

图5-1　激光雷达

激光雷达的优点是测量精确，能够比较精准地提供角度和距离信息，可以达到小于1°的角度精度以及cm级别的测距精度，扫描范围广（通常能够覆盖平面内270°以上的范围）。而且基于扫描振镜式的固态激光雷达（如Sick、Hokuyo等）可以达到较高的数据刷新率（20Hz以上），基本满足了实时操作的需要，其缺点是价格比较昂贵（目前市面上比较便宜的机械旋转式单线激光雷达也要几千元），安装部署对结构有要求（要求扫描平面无遮挡）。

激光雷达SLAM建立的地图常常使用占据栅格地图（Ocupanccy Grid）表示，

每个栅格以概率的形式表示被占据的概率，存储非常紧凑，特别适合于进行路径规划，如图 5-2 所示。

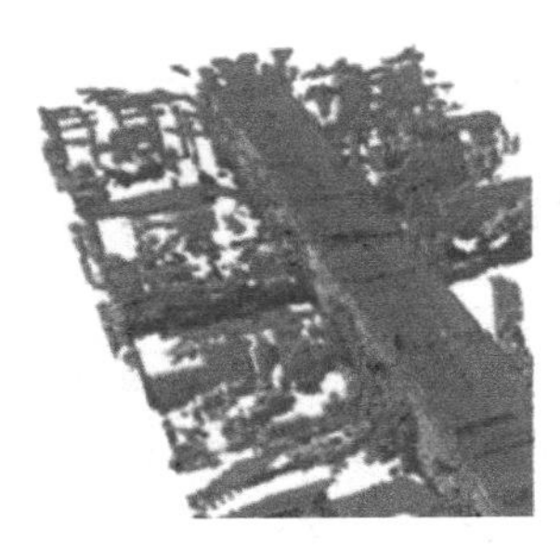

图 5-2 占据栅格地图

经典著作 *Probabilistic Robotics* 一书中详细阐述了利用 2D 激光雷达基于概率方法进行地图构建和定位的理论基础，并阐述了基于 RBPF 粒子滤波器的 FastSLAM 方法，成为后来 2D 激光雷达建图的标准方法之一 gmapping1 的基础，该算法也被集成到机器人操作系统（Robot Operation System，ROS）中，这是我们下一节将要介绍并使用的方法。

2016 年，Google 开源其激光雷达 SLAM 算法库 Cartographer4，改进了 gmapping 计算复杂、不能有效处理闭环的缺点，采用 SubMap 和 Scan Match 的思想构建地图，能够有效地处理闭环，达到了较好的效果。

5.2 gmapping建图功能应用

gmapping 是 ROS 中自带的建图功能包，安装 ROS 时已经默认安装 gmapping。若未安装，可以通过以下命名安装（以 ROS indigo 版本为例）：

```
sudo apt-get install ros-indigo-slam-gmapping
```

在运行建图功能包前，应提供激光雷达传感器的数据及机器人运动变换关系 TF。这两个数据是 gmapping 接收的消息，消息如下。

（1）tf (tf/tfMessage)

（2）scan (sensor_msgs/LaserScan)

发布的地图信息消息如下。

（1）map_metadata (nav_msgs/MapMetaData)

（2）map (nav_msgs/OccupancyGrid)

在运行建图功能包前，我们通常需要一个具备激光传感器的机器人及其工作环境。或者，在硬件环境不允许的情况下，你至少应该有一个具备类似功能的仿真软件环境（在后续章节，我们会介绍如何搭建机器人仿真环境）。现在，即使你还不具备这些，也并不妨碍你体验 gmapping，因为我们可以使用网上已有的激光数据资源。接下来，先下载激光数据集到本地。下载地址为 https://github.com/RobInLabUJI/ROS-Stage/blob/master/basic_localization_stage.bag。

启动 ROS 节点管理器：

```
roscore
```

启用重放时的数据中的时间，不是现在的本地时间，而是历史时间。因此我们需要告诉 ROS 系统，使用历史模拟时间：

```
rosparam set use_sim_time true
```

运行 slam_gmapping 功能包，并设置激光雷达的 topic 为 base_scan：

```
rosrun gmapping slam_gmapping scan:=base_scan
```

通过 rosbag 回放激光雷达数据：

```
cd pathtoyourbagfile(转到存放下载激光数据的文件目录下)
rosbag play --clock basic_localization_stage.bag
```

运行 Rviz，并选择 /map 和 /scan 的 topic：

```
rosrun rviz rviz
```

可以看到图 5-3 所示的建图效果。

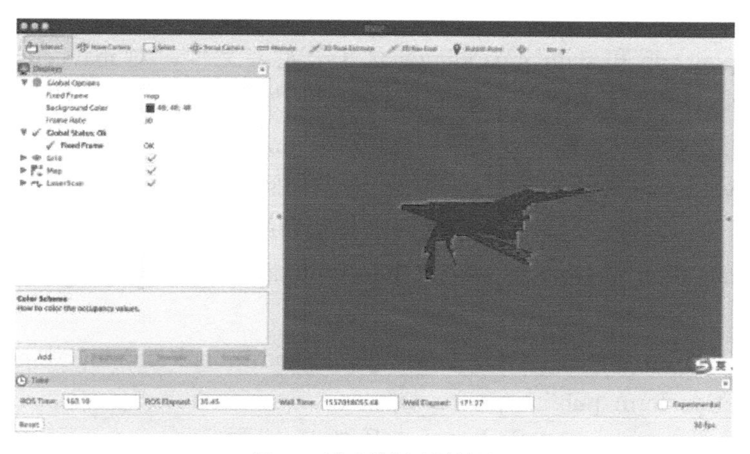

图 5-3　激光数据建图效果

5.3 ROS gmapping功能包解读

ROS gmapping 功能包是基于 openslam 开源功能包实现的，是对 openslam 的封装。其主要功能粒子滤波和 scanmatch 等都是在 openslam 中完成的。现在，我们先不讨论如何实现这些功能，先看看 ROS gmapping 功能包的主要逻辑结构。

下载 ros hydro-devel 版本的 gmapping 源码，下载地址为 https://github.com/ros-perception/slam_gmapping。

你可以选择任何文本编辑软件查看该源码，为了方便跳转，可以安装 understand 软件进行阅读（请读者根据实际需求安装，不安装也不影响），安装步骤如下。

进入 understand 官网下载对应的版本（32 位或 64 位），本书以 64 位安装为例。

（1）解压到安装目录：

```
gzip -cd Understand-3.1.677-Linux-64bit.tgz  | tar xvf -
```

（2）添加路径：

```
export PATH=$PATH:/home/myname/scitools/bin/linux64
export STIHOME=/home/myname/scitools
```

（3）安装：

```
cd pathtounderstand/scitools/bin/linux64 && ./understand
```

用 understand 软件打开 gmapping 源码包，在 src 文件夹下可看到 main.cpp、slam_gmapping.cpp 和 slam_gmapping.h 这 3 个主要文件，对 main.cpp 运行软件上的 Butterfly，可以看到整个工程的调用关系，如图 5-4 所示。

主函数通过 SlamGMapping 类实例化类对象 gn，然后进入 startLiveSlam() 函数。

startLiveSlam() 函数主要完成了一些消息订阅与发布（如激光数据、地图数据）等初始化的功能。这里最主要的两点分别为：

（1）scan_filter_->registerCallback(boost::bind(&SlamGMapping::laserCallback, this, _1));

（2）transform_thread_ = new boost::thread(boost::bind(&SlamGMapping::publishLoop, this, transform_publish_period_));

其中，scan_filter_ 用于完成激光雷达数据的处理，transform_thread_ 用于发布机器人里程计等 TF 坐标变换。

激光雷达的数据处理通过 laserCallback() 函数实现，读者可以使用 Butterfly 功能查看该函数的调用关系，如图 5-5 所示。

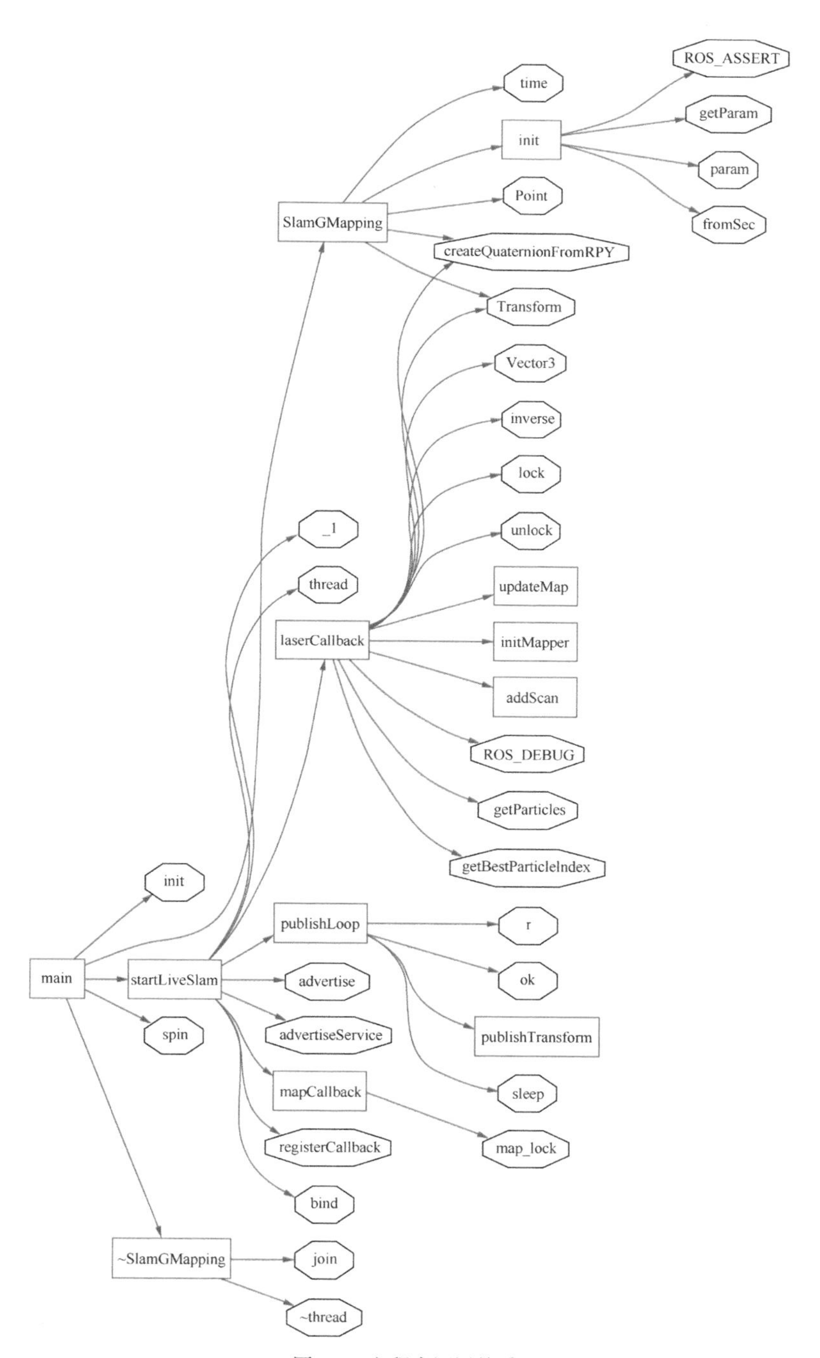

ROS_ASSERT
time
getParam
init
param
Point
SlamGMapping
fromSec
createQuaternionFromRPY
Transform
Vector3
inverse
lock
_1
unlock
thread
updateMap
laserCallback
initMapper
addScan
ROS_DEBUG
getParticles
getBestParticleIndex
init
publishLoop
r
main
startLiveSlam
advertise
ok
spin
advertiseService
publishTransform
mapCallback
sleep
registerCallback
map_lock
bind
~SlamGMapping
join
~thread

图 5-4　主程序调用关系

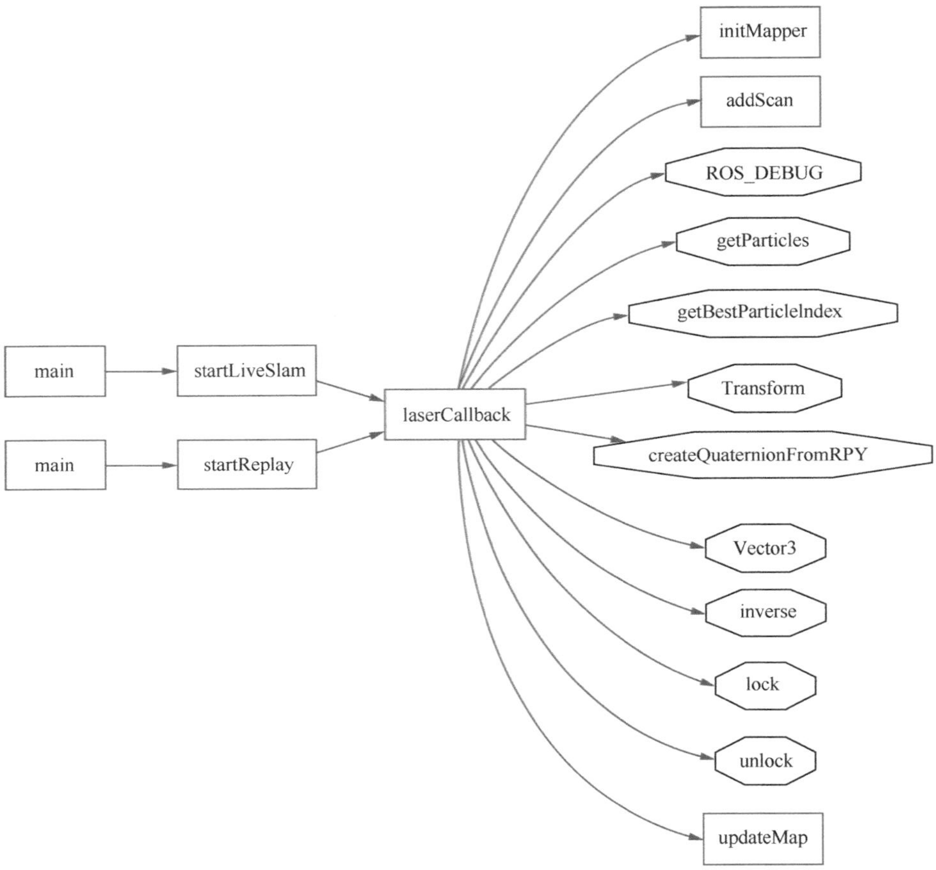

图 5-5 laserCallback() 函数调用关系

laserCallback() 函数对激光雷达的处理主要通过 addScan 函数实现。该函数的定义原型如下：

```
SlamGMapping::addScan(const sensor_msgs::LaserScan& scan, GMapping::OrientedPoint& gmap_pose)
```

其中，gmap_pose 是机器人里程计坐标。接下来通过 gsp_->processScan(reading) 进入 openslam 对激光数据进行处理。无论 processScan 函数如何复杂，其功能就是通过激光数据来辅助机器人定位，进而获取更准确的机器人位置坐标。

当获取到机器人的位置坐标后，我们就可以根据该坐标及激光雷达数据更新地图了。注意：此处的更新地图是通过获取最优粒子，进而得到该粒子所携带的地图（openslam 中每个粒子携带一幅地图）。这部分功能的实现是通过 updateMap(*scan) 实现的，该函数的定义如下：

```
SlamGMapping::updateMap(const sensor_msgs::LaserScan& scan)
```

获取最优粒子的代码如下：

```
GMapping::GridSlamProcessor::Particle best =
                                gsp_->getParticles()[gsp_->getBestParticleIndex()];
```

根据最优粒子获取机器人轨迹上的点，并获取地图数据：

```
for(GMapping::GridSlamProcessor::TNode* n = best.node;
    n;
    n = n->parent)
{
  ROS_DEBUG("  %.3f %.3f %.3f",
            n->pose.x,
            n->pose.y,
            n->pose.theta);
  if(!n->reading)
  {
    ROS_DEBUG("Reading is NULL");
    continue;
  }
  matcher.invalidateActiveArea();
  matcher.computeActiveArea(smap, n->pose, &((*n->reading)[0]));
  matcher.registerScan(smap, n->pose, &((*n->reading)[0]));
}
```

将更新的地图数据封装成 ROS 的 map 消息：

```
for(int x=0; x < smap.getMapSizeX(); x++)
{
  for(int y=0; y < smap.getMapSizeY(); y++)
  {
    /// @todo Sort out the unknown vs. free vs. obstacle thresholding
    GMapping::IntPoint p(x, y);
    double occ=smap.cell(p);
    assert(occ <= 1.0);
    if(occ < 0)
```

```
        map_.map.data[MAP_IDX(map_.map.info.width, x, y)] = -1;
      else if(occ > occ_thresh_)
      {
        //map_.map.data[MAP_IDX(map_.map.info.width, x, y)] = (int)round(occ*100.0);
        map_.map.data[MAP_IDX(map_.map.info.width, x, y)] = 100;
      }
      else
        map_.map.data[MAP_IDX(map_.map.info.width, x, y)] = 0;
    }
  }
```

至此，我们就完成了 ROS gmapping 建图的源码解读。

5.4 openslam源码解读

ROS gmapping 功能包是基于 openslam 实现的，本节我们来学习 openslam 源码。openslam 是基于 Rao-Blackwellized 粒子滤波器的建图功能包。有关 openslam 的介绍可在其官网上查阅，官网中还包含 orbslam 等其他优秀的开源项目，有兴趣的读者可根据自身需求进行学习，官网地址为 https://openslam-org.github.io/。

由于 ROS 的 gmapping 功能包是基于 openslam_gmapping 实现的，因此本节我们来研究 openslam_gmapping。下载 openslam_gmapping 的源码，下载地址为 https://github.com/OpenSLAM-org/openslam_gmapping。

下载之后，使用 understand 软件打开。主文件是在 gui 文件下的 gfs_simplegui.cpp，打开之后进行 Butterfly，可以看到调用关系如图 5-6 所示。

在上述调用关系中，两个 start 分别对应 gfs_simplegui.cpp 中的如下两个方法：

```
gsp->start();
mainWin->start(1000);
```

这里主要关注 gsp->start()，因为所有的处理过程都在此方法中完成。对 gsp->start() 运行 Butterfly，查看其调用关系如图 5-7 所示。

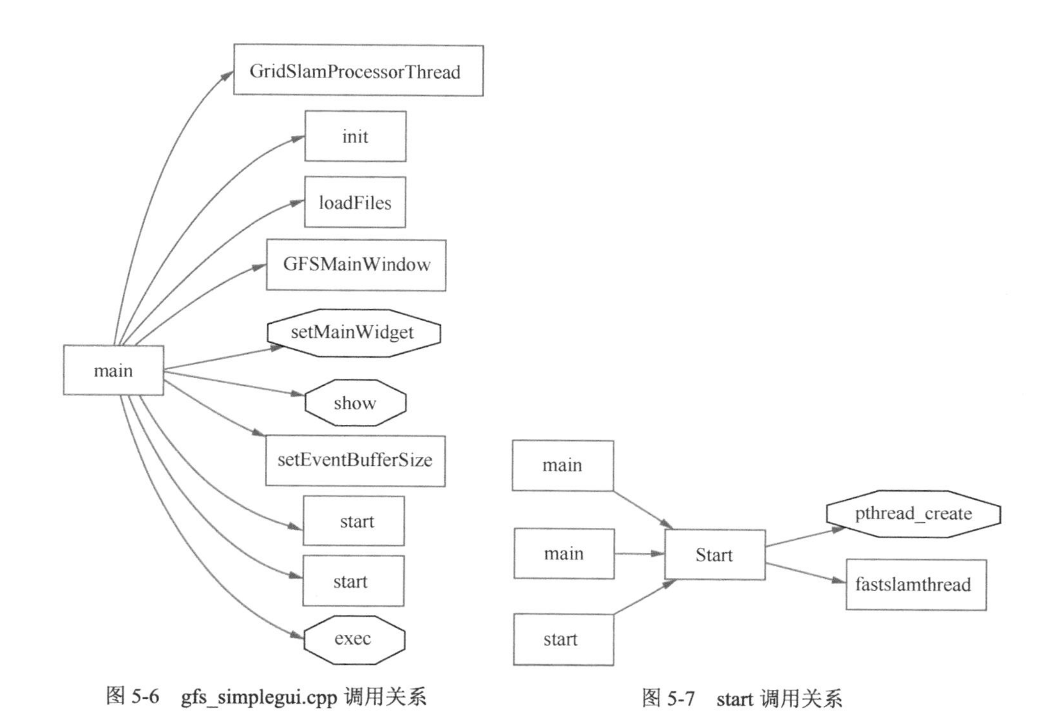

图 5-6 gfs_simplegui.cpp 调用关系

图 5-7 start 调用关系

查看 fastslamthread 调用关系，如图 5-8 所示。

可以看到，ROS gampping 中调用的 processScan() 函数已经出现了，查看该函数的调用关系，如图 5-9 所示。

可以看到 processScan 主要调用了 getPose、scanMatch、drawFrom Motion、resample、updateTreeWeights 这些函数。

接下来，我们依次了解这些函数的功能。首先，获取里程计坐标，通过 getPose() 实现。

```
OrientedPoint relPose=reading.getPose();
```

然后通过 drawFromMotion 更新所有粒子的位置坐标值：

```
pose=m_motionModel.drawFromMotion(it->pose, relPose, m_odoPose);
```

获取机器人相邻两个时刻的位移（包含旋转和平移）：

```
OrientedPoint move=relPose-m_odoPose;
```

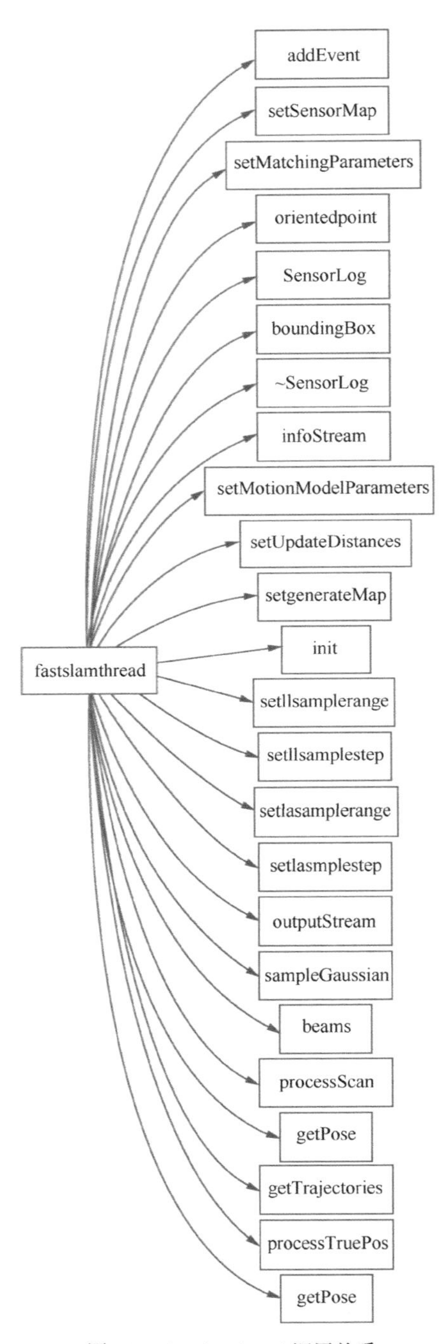

图 5-8　fastslamthread 调用关系

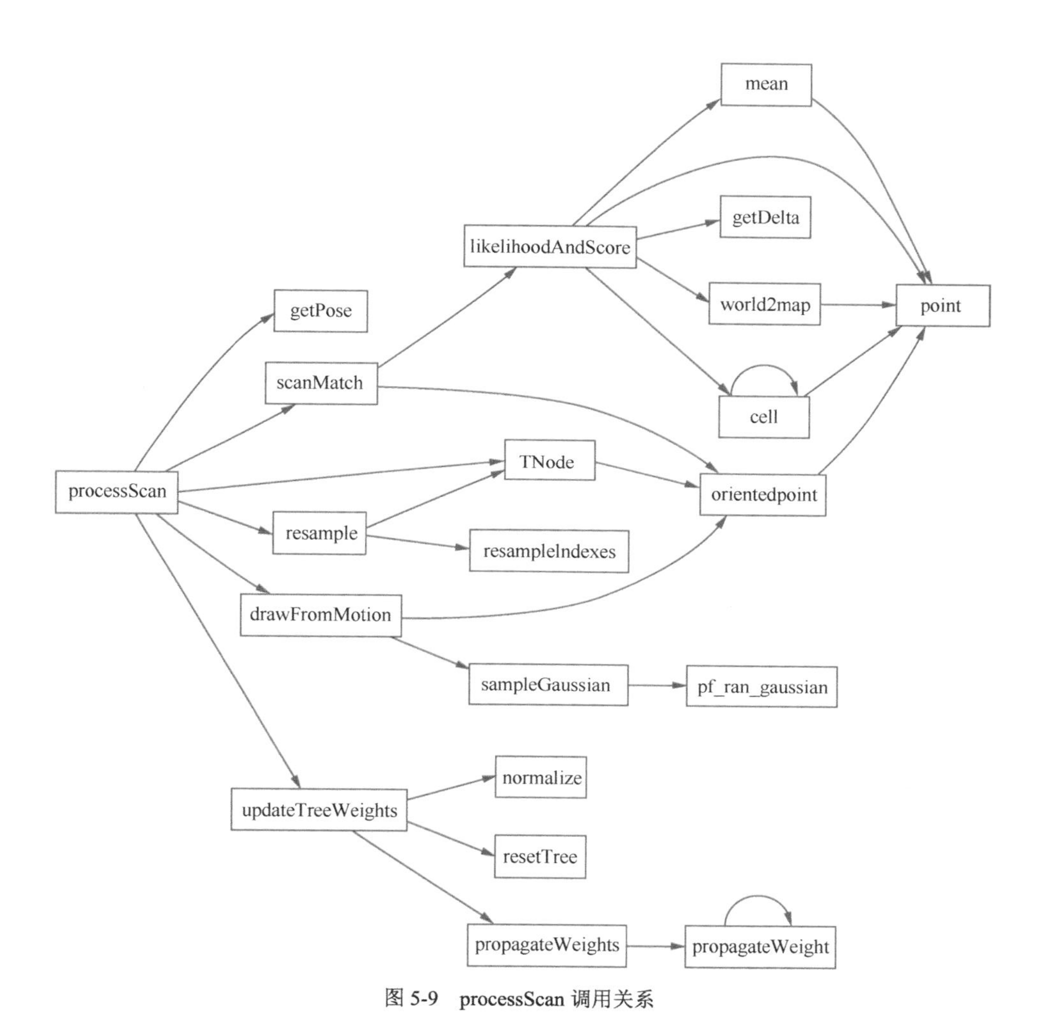

图 5-9 processScan 调用关系

通过位移判断机器人运动是否存在跳变：

```
if (m_linearDistance>m_distanceThresholdCheck)
```

判断机器人是否移动足够的距离或旋转足够的角度，仅当条件满足时才处理激光数据：

```
    if (! m_count
    || m_linearDistance>m_linearThresholdDistance
    || m_angularDistance>m_angularThresholdDistance)
```

复制激光雷达数据，对复制的数据进行处理：

```
double * plainReading = new double[m_beams];
for(unsigned int i=0; i<m_beams; i++){
  plainReading[i]=reading[i];
}
```

通过 scanMatch(plainReading) 处理复制的激光雷达数据。scanMatch 函数在 gridslamprocessor.hxx 文件中定义，函数原型如下：

```
inline void GridSlamProcessor::scanMatch(const double* plainReading)
```

该函数的调用关系如图 5-10 所示。

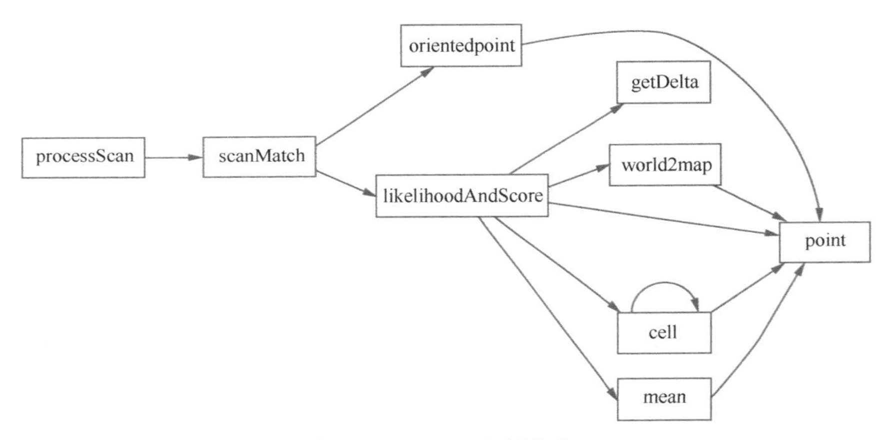

图 5-10 scanMatch 调用关系

在 scanMatch 函数中，主要通过：

```
score=m_matcher.optimize(corrected, it->map, it->pose, plainReading);
```

获取每个粒子的最优位姿，优化的方法是前后左右微小地移动粒子，获取最好的匹配效果，将其对应的姿态作为该粒子的位姿，然后计算该粒子的权重：

```
m_matcher.likelihoodAndScore(s, l, it->map, it->pose, plainReading);
sumScore+=score;
it->weight+=l;
it->weightSum+=l;
```

更新粒子携带的地图：

```
m_matcher.invalidateActiveArea();
m_matcher.computeActiveArea(it->map, it->pose, plainReading);
```

更新权重：

```
updateTreeWeights(false);
```

注意：参数 false 代表是否进行归一化。函数原型如下：

```
void  GridSlamProcessor::updateTreeWeights(bool weightsAlreadyNormalized){
  if (!weightsAlreadyNormalized) {
    normalize();
  }
  resetTree();
  propagateWeights();
}
```

重采样，防止粒子退化：

```
resample(plainReading, adaptParticles);
```

更新数据，为下一次激光数据处理做准备：

```
    delete [] plainReading;
    m_lastPartPose=m_odoPose; //update the past pose for the next iteration
    m_linearDistance=0;
    m_angularDistance=0;
    m_count++;
    processed=true;

    //keep ready for the next step
    for (ParticleVector::iterator it=m_particles.begin(); it!=m_particles.
end(); it++){
  it->previousPose=it->pose;
    }
```

至此，processScan() 函数已经梳理完毕。然后，进入循环处理流程，每次激光雷达数据获取之后就进入该函数进行处理。

> **注意**
>
> 每个粒子携带一幅地图，最后选取的是最优的地图。在处理地图数据时，openslam 使用了 bresenham 算法。该算法源码在 openslam 源码包的 scanmatcher 文件夹下的 gridlinetraversal.h 中，有兴趣的读者可以自己查阅其实现过程。

下面我们简单介绍一些算法，并编写一个小程序实现其直线可视化过程。

bresenham's line algorithm 即布雷森汉姆直线算法，用于描绘由两点所决定的直线，它会算出一条线段在 n 维位图上最接近的点。这个算法只会用到较为快速的整数加法、减法和位元移位，常用于绘制电脑画面中的直线，是计算机图形学中最先发展出来的算法。

假设由 (x_0, y_0) 这一点，绘画一直线至右下角的另一点 (x_1, y_1)，x、y 分别代表其水平及垂直坐标，并且 $x_1-x_0 > y_1-y_0$。在此我们使用电脑系统常用的坐标系，即 x 坐标值沿 x 轴向右增长，y 坐标值沿 y 轴向下增长，如图 5-11 所示。

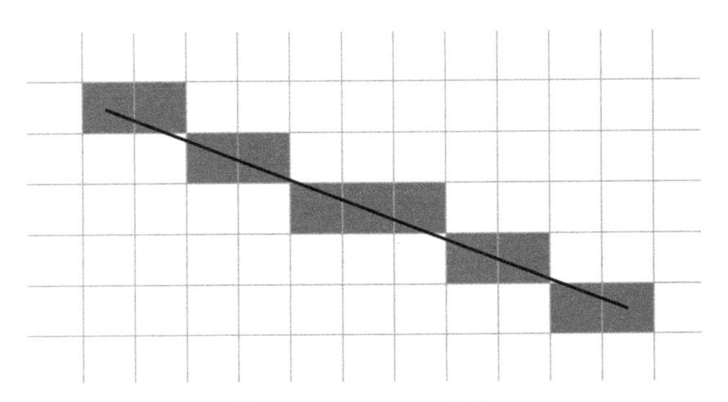

图 5-11　bresenham 画线

由 (x_0, y_0) 及 (x_1, y_1) 两点所组成的直线，其公式如下：

$$y - y_0 = \frac{y_1 - y_0}{x_1 - x_0}(x - x_0)$$

因此，对于每一点的 x，其 y 的值是：

$$\frac{y_1 - y_0}{x_1 - x_0}(x - x_0) + y_0$$

在实际运算中，我们需要计算每一像素点与该线之间的误差。误差为在每一点 x 中，其相对的像素点的 y 值与该线实际的 y 值的差距。每当 x 的值增加 1，误差的值就会增加 m。每当误差的值超出 0.5，线就会比较靠近下一个映像点，因此 y 的值便会加 1，且误差减 1。此外，通过交换 x、y 并使 y 值增加 -1 等操作可以实现负斜率及任意方向直线画法。算法的伪代码如下：

```
function line(x0, x1, y0, y1)
    boolean steep := abs(y1 - y0) > abs(x1 - x0)
    if steep then
        swap(x0, y0)
        swap(x1, y1)
    if x0 > x1 then
        swap(x0, x1)
        swap(y0, y1)
```

```
    int deltax := x1 - x0
    int deltay := abs(y1 - y0)
    real error := 0
    real deltaerr := deltay / deltax
    int ystep
    int y := y0
    if y0 < y1 then ystep := 1 else ystep := -1
    for x from x0 to x1
        if steep then plot(y,x) else plot(x,y)
        error := error + deltaerr
        if error ≥ 0.5 then
            y := y + ystep
            error := error - 1.0
```

基于上述伪代码，在Ubuntu系统上利用OPenGl对该算法进行可视化实现。实现代码如下面的main.cpp所示，读者可以通过code::blocks等IDE打开该main.cpp进行测试。

```
#include <iostream>
#include <GL/glut.h>
#include <vector>
using namespace std;
int x0 = 500,y0 = 0;
int x1 = -500,y1 =500;
//define points
struct MapPoint
{
    int x;
    int y;
    MapPoint()
    {
        x=0;
        y=0;
    }
};

//draw points
void putpixel(int x,int y)
{
    glColor3f(0.0, 1.0, 1.0);
    glPointSize(3.0);
    glBegin(GL_POINTS);
    glVertex2f(x, y);
    glEnd();
```

```
}
//used for draw line
void bresenham_line(int x0, int y0, int x1, int y1) {
    int dx, dy, h, a, b, x, y, flag, t;
    dx = abs(x1 - x0);
    dy = abs(y1 - y0);
    if (x1 > x0) a = 1; else a = -1;
    if (y1 > y0) b = 1; else b = -1;
    x = x0;
    y = y0;
    if (dx >= dy) {
        //0< |k| <=1
        flag = 0;
    } else {
        //|k|>1,exchange dx, dy
        t = dx;
        dx = dy;
        dy = t;
        flag = 1;
    }
    h = 2 * dy - dx;
    for (int i = 1; i <= dx; ++i) {
        putpixel(x, y);
        if (h >= 0) {
            if (flag == 0) y = y + b;
            else x = x + a;
            h = h - 2 * dx;
        }
        if (flag == 0) x = x + a;
        else y = y + b;
        h = h + 2 * dy;

    }

}

//used for obtain points set
vector<MapPoint> bresenham(int x0,int y0,int x1,int y1)
{
    vector<MapPoint> pp;
    MapPoint p;
    cout<<"init p is: "<<p.x<<" "<<p.y<<endl;
    int dx, dy, h, a, b, x, y, flag, t;
    dx = abs(x1 - x0);
```

```
    dy = abs(y1 - y0);
    if (x1 > x0) a = 1; else a = -1;
    if (y1 > y0) b = 1; else b = -1;
    x = x0;
    y = y0;
    if (dx >= dy) {
        //0< |k| <=1
        flag = 0;
    } else {
        //|k|>1,exchange dx, dy
        t = dx;
        dx = dy;
        dy = t;
        flag = 1;
    }
    h = 2 * dy - dx;
    for (int i = 1; i <= dx; ++i) {
        p.x = x,p.y=y;
        pp.push_back(p);
        if (h >= 0) {
            if (flag == 0) y = y + b;
            else x = x + a;
            h = h - 2 * dx;
        }
        if (flag == 0) x = x + a;
        else y = y + b;
        h = h + 2 * dy;

    }
    return pp;
}

//gult init
void init()
{
glutInitDisplayMode(GLUT_SINGLE|GLUT_RGB);
glutInitWindowPosition(0,0);
glutInitWindowSize(600,600);
glutCreateWindow("Bresenham");
    glClear(GL_COLOR_BUFFER_BIT);
glClearColor(0.0, 0.0, 0.0, 1.0);
gluOrtho2D(-500, 500, -500, 500);
}
```

```
//show result
void display()
{
    bresenham_line(x0,y0,x1,y1);
    glFlush();
}

int main(int argc,char** argv)
{
glutInit(&argc, argv);
init();
vector<MapPoint> pset;
pset = bresenham(x0,y0,x1,y1);
// print points in pset
for(vector<MapPoint>::iterator iter=pset.begin();
                               iter!=pset.end();iter++)
{
        cout<<(*iter).x<<" "<<(*iter).y<<endl;
}
//display draw points
    glutDisplayFunc(display);
glutMainLoop();
    return 0;
}
```

在本程序中，读者可通过修改 x_0、x_1、y_0、y_1 的值来测试画不同直线的效果。本例的测试效果如图 5-12 所示。

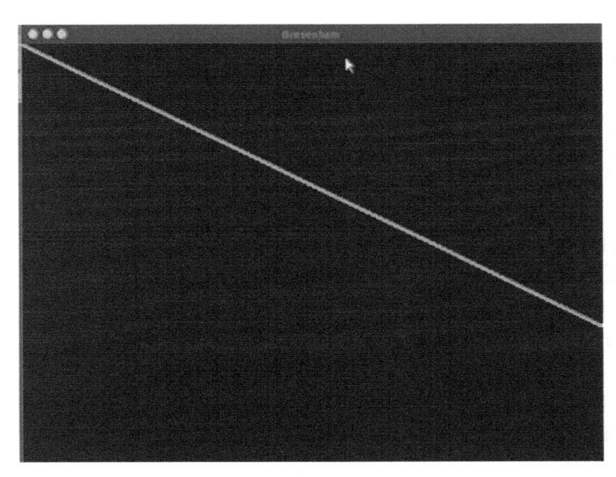

图 5-12　画线效果可视化

5.5 ROS建图实战

经过前面内容的铺垫，相信读者对SLAM建图已经有了相对清晰的认识。但是，仅使用封装好的功能包对于深刻理解SLAM建图的意义并不大。接下来，我们通过一个简单的实例来创建自己的建图功能包，从而提高读者对建图原理的理解。

5.5.1 ROS地图发布

ROS中map的消息格式主要有nav_msgs::OccupancyGrid和nav_msgs::MapMetaData两种。其中，nav_msgs::MapMetaData定义了地图的原点和地图长宽等信息；nav_msgs::OccupancyGrid则包含了nav_msgs::MapMetaData消息。使用rosmsg查看消息具体内容的方法如下：

```
rosmsg show  nav_msgs/MapMetaData
time map_load_time
float32 resolution
uint32 width
uint32 height
geometry_msgs/Pose origin
  geometry_msgs/Point position
    float64 x
    float64 y
    float64 z
  geometry_msgs/Quaternion orientation
    float64 x
    float64 y
    float64 z
    float64 w
rosmsg show nav_msgs/OccupancyGrid
std_msgs/Header header
  uint32 seq
  time stamp
  string frame_id
nav_msgs/MapMetaData info
  time map_load_time
  float32 resolution
```

```
    uint32 width
    uint32 height
    geometry_msgs/Pose origin
      geometry_msgs/Point position
        float64 x
        float64 y
        float64 z
      geometry_msgs/Quaternion orientation
        float64 x
        float64 y
        float64 z
        float64 w
  int8[] data
```

其中，data 中包含的是地图中每个点的数据信息。

接下来，我们自行填充 map 数据进行发布，首先创建 book_mappub.h 头文件，内容如下：

```
#include "ros/ros.h"
#include "nav_msgs/OccupancyGrid.h"
#include "nav_msgs/MapMetaData.h"

class MapPub
{
    public:
        MapPub(ros::NodeHandle& nh,double map_reso,
                int map_width,int map_height,
                double posx,double posy,double posz,
                double orientx,double orienty,
                double orientz,double orientw);
        ~MapPub();
        void mappub_init();
        void mapdata_init();
    private:
        ros::NodeHandle nh_;
        ros::Publisher mappub_;
        nav_msgs::OccupancyGrid mapdata_;
        nav_msgs::MapMetaData datainfo_;
        double map_reso;
        int map_width,map_height;
        double posx,posy,posz;
        double orientx,orienty,orientz,orientw;
};
```

上述代码中主要定义了 MapPub 类及一些地图数据信息。然后创建 book_mappub.cpp 文件，实现具体功能，代码内容如下：

```
#include "book_mappub.h"
MapPub::MapPub(ros::NodeHandle& nh,double map_reso,
                int map_width,int map_height,
                double posx,double posy,double posz,
                double orientx,double orienty,
                double orientz,double orientw):
                nh_(nh),map_reso(map_reso),map_width(map_width),
                map_height(map_height),posx(posx),posy(posy),posz(posz),
                orientx(orientx),orienty(orienty),orientz(orientz),
                orientw(orientw)
{
    mapdata_init();
    mappub_init();
}

MapPub::~MapPub()
{
}

void MapPub::mapdata_init()
{
    //nav_msgs::MapMetaData datainfo_ fill
    ros::Time maptime = ros::Time::now();
    datainfo_.resolution = map_reso;
    datainfo_.width = map_width;
    datainfo_.height = map_height;
    datainfo_.origin.position.x = posx;
    datainfo_.origin.position.y = posy;
    datainfo_.origin.position.z = posz;
    datainfo_.origin.orientation.x = orientx;
    datainfo_.origin.orientation.y = orienty;
    datainfo_.origin.orientation.z = orientz;
    datainfo_.origin.orientation.w = orientw;

    int datasize = map_width*map_height;
    // nav_msgs::OccupancyGrid mapdata_ fill
    mapdata_.header.stamp = maptime;
    mapdata_.header.frame_id = "map";
    mapdata_.info = datainfo_;
    mapdata_.data.resize(datasize);
```

```
    int obsx1_min = map_height/3*2*map_width + map_width/4;
    int obsx1_max = map_height/3*2*map_width + map_width;
    int obsx2_min = map_height/3*map_width;
    int obsx2_max = map_height/3*map_width + map_width/3;
    int obsx3_min = (map_height-1)*map_width;
    int obsx3_max = map_height*map_width;
    int obsy_max = map_width*map_height/2+map_width/3*2;
    int obsy = map_width/3*2;
    //mapdata_ data fill
    for(int i=0;i<datasize;i++)
    {
        int flag = i%map_width;

        if((i>=0 && i<=map_width) || (i>=obsx3_min && i<=obsx3_max) ||
           (i>=obsx1_min && i<=obsx1_max) || (i>=obsx2_min && i<=obsx2_max))
        {
            mapdata_.data[i] = 100;
        }else if(flag==0 || flag==(map_height-1) || (flag==obsy && i<=obsy_max))
        {
            mapdata_.data[i] = 100;
        }
        else{
            mapdata_.data[i] = 0;
        }
    }
}

void MapPub::mappub_init()
{
    ros::Rate rate(10);
    // map publisher initializer
    mappub_ = nh_.advertise<nav_msgs::OccupancyGrid>("map",1000);
    while(nh_.ok())
    {
        // publish map topic
        mappub_.publish(mapdata_);
        rate.sleep();
    }
}

int main(int argc,char** argv)
{
    //init and name node
```

```
    ros::init(argc, argv, "mappub_demo");
    //create ros node handle
    ros::NodeHandle nh_;
    // map param info
    double resol = 0.05;
    int mapw = 200,maph = 200;
    double psx=0,psy=0,psz=0;
    double rx=0,ry=0,rz=0,rw=1;
    //MapPub instantiate object
    MapPub mappub(nh_,resol,mapw,maph,psx,psy,psz,rx,ry,rz,rw);
    return 0;
}
```

说明如下。

- void MapPub::mapdata_init()：实现地图数据初始化填充，主要包含地图原点、长度、宽度及各点的状态值。主要填充过程在 for(int i=0;i<datasize;i++) 循环结构中。
- void MapPub::mappub_init()：实现地图数据发布。
- mappub_ = nh_.advertise<nav_msgs::OccupancyGrid>("map",1000)：创建发布者，并在“map”上发布地图数据。
- while(nh_.ok())：在该循环体中实现地图循环发布，发布频率 10Hz。

创建 CMakeLists.txt 文件，内容如下：

```
cmake_minimum_required(VERSION 2.8.3)
project(book_mappub)
find_package(catkin REQUIRED COMPONENTS roscpp rospy std_msgs nav_msgs)
catkin_package(
#  INCLUDE_DIRS include
#  LIBRARIES book_mappub
#  CATKIN_DEPENDS other_catkin_pkg
#  DEPENDS system_lib
)

## Specify additional locations of header files
## Your package locations should be listed before other locations
include_directories(
  include ${catkin_INCLUDE_DIRS}
# include
# ${catkin_INCLUDE_DIRS}
)
```

```
    add_executable(book_mappub
      src/book_mappub.cpp
    )
    add_dependencies(book_mappub ${${PROJECT_NAME}_EXPORTED_TARGETS} ${catkin_
EXPORTED_TARGETS})
    target_link_libraries(book_mappub
      ${catkin_LIBRARIES}
    )
```

测试程序，查看地图形状：

```
rosrun book_mappub book_mappub
rosrun rivz rviz
```

选择 map topic，可以看到地图如图 5-13 所示。

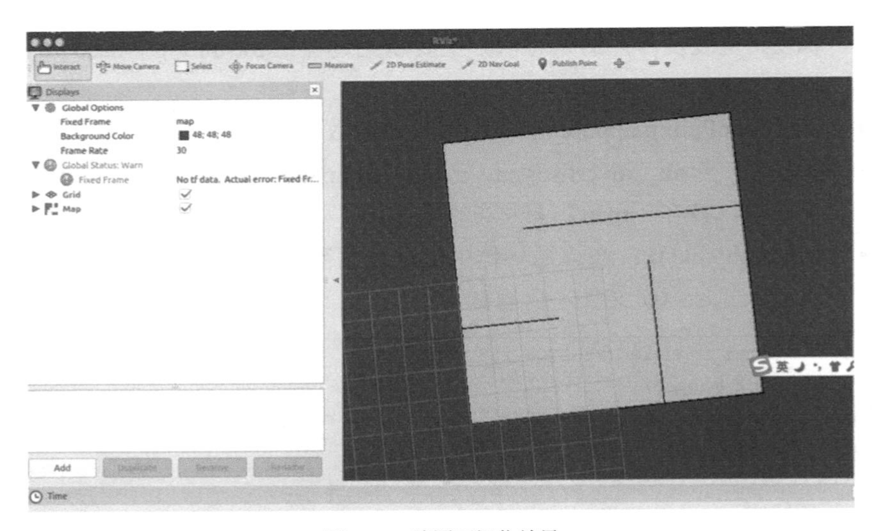

图 5-13　地图可视化效果

5.5.2　TF坐标变换发布

在建图过程中，各坐标系的变换关系必不可少。在本节中，我们将介绍如何发布机器人 base_link 坐标系与地图 map 坐标系的关系（将激光雷达坐标系设置为与机器人 base_link 坐标系重合）。首先，创建 slam_tfpub.h 头文件，内容如下：

```
#include "ros/ros.h"
#include "geometry_msgs/Twist.h"
```

```
#include "tf/transform_broadcaster.h"
#include "tf/tf.h"
#define pi 3.1415926

class TfMove
{
    public:
        TfMove(ros::NodeHandle& nh,ros::Rate& r);
        void VelCallback(const geometry_msgs::TwistPtr& vel);
        void init_sub();
    private:
        ros::NodeHandle nh_;
        ros::Subscriber sub_;
        tf::TransformBroadcaster tfbrd_;
        ros::Rate rate;
        double x,y,z,roll,pit,yaw;
};
```

接下来，创建 slam_tfpub.cpp 文件完成具体方法的实现，代码内容如下：

```
#include "slam_tfpub.h"
//TfMove initialize
TfMove::TfMove(ros::NodeHandle& nh,ros::Rate& r):nh_(nh),rate®,
                                                  x(0),y(0),z(0),
                                                  roll(0),
                                                  pit(0),
                                                  yaw(0)
{
    init_sub();
}

//velocity sub callback and send tansform between odom and base_link
void TfMove::VelCallback(const geometry_msgs::TwistPtr& vel)
{
    // linear change
    x = x+vel->linear.x;
    y = y+vel->linear.y;
    z = z+vel->linear.z;
    //angular change
    roll = roll+vel->angular.x/pi*180;
    pit = pit+vel->angular.y/pi*180;
    yaw = yaw+vel->angular.z/pi*180;
    // transform between odom and base_link
    tf::Transform trans;
```

```
        trans.setOrigin(tf::Vector3(x,y,z));
        tf::Quaternion q;
        q.setRPY(roll,pit,yaw);
        trans.setRotation(q);
        //send transform  map --- parent frame , base_link --- child frame
        tfbrd_.sendTransform(tf::StampedTransform(trans,ros::Time::now(),"map","
base_link"));
        rate.sleep();
    }

    //initial sub
    void TfMove::init_sub()
    {
        sub_ = nh_.subscribe("cmd_vel",1,&TfMove::VelCallback,this);
        ros::spin();
    }

    int main(int argc,char** argv)
    {
        //initialize and name mode
        ros::init(argc,argv,"myslam_tfpub");
        //create node handle
        ros::NodeHandle nh;
        // sleep rate
        ros::Rate rate(10);
        //TfMove object instantiate
        TfMove tfmove(nh,rate);
        return 0;
    }
```

说明如下。

- void TfMove::VelCallback(const geometry_msgs::TwistPtr& vel) 定义了 TF 从外部接收 geometry_msgs::TwistPtr 消息，并根据此内容确定 base_link 与 map 的坐标变换关系。
- x,y,z,roll,pit,yaw 值的变化表示 base_link 的原点在 map 坐标系中的坐标。
- tfbrd_.sendTransform(tf::StampedTransform(trans,ros::Time::now(),"map","base_link")) 用于发布坐标变换关系。

5.5.3 模拟激光数据

除了 TF 坐标变换关系，激光雷达数据也是必不可少的。许多读者手中可能并没有昂贵的激光雷达设备，但这并不影响我们体验建图原理。接下来，我们利用 ROS 中 scan 的消息格式伪造激光数据。

创建 slam_laser.h 头文件，内容如下：

```
#include "ros/ros.h"
#include "sensor_msgs/LaserScan.h"
class LaserScanPub
{
    public:
        LaserScanPub(ros::NodeHandle& nh,
                     double minAngle,double maxAngle,double scanTime,
                     double minRange,double maxRange,double scanNums);
        ~LaserScanPub();
        void scanpub_init();
        void laserdata_init();
    private:
        ros::NodeHandle nh_;
        ros::Publisher scanpub_;
        sensor_msgs::LaserScan laserdata_;
        double minAngle,maxAngle;
        double minRange,maxRange;
        double scanTime,scanNums;
};
```

这里主要定义了激光雷达的数据对象，以及扫描距离和角度等信息。然后创建 slam_laser.cpp 文件，实现具体方法，代码内容如下：

```
#include "slam_laser.h"

//LaserScanPub initialize
LaserScanPub::LaserScanPub(ros::NodeHandle& nh,
                    double minAngle,double maxAngle,double scanTime,
                    double minRange,double maxRange,double scanNums):
            nh_(nh),minAngle(minAngle),maxAngle(maxAngle),minRange(minRange),
            maxRange(maxRange),scanNums(scanNums),scanTime(scanTime)
{
    //laserdata_init();
    scanpub_init();
}

LaserScanPub::~LaserScanPub()
{
}
```

```
//laserdata init
void LaserScanPub::laserdata_init()
{
    //create laser data
    ros::Time scantime = ros::Time::now();
    laserdata_.header.stamp = scantime;
    laserdata_.header.frame_id = "base_link";
    laserdata_.range_min = minRange;
    laserdata_.range_max = maxRange;
    laserdata_.scan_time = scanTime;
    laserdata_.angle_increment = (maxAngle-minAngle)/scanNums;
    laserdata_.time_increment = scanTime/scanNums;
    laserdata_.ranges.resize(scanNums);
    laserdata_.intensities.resize(scanNums);
    for(int i=0;i<scanNums;i++)
    {
        laserdata_.ranges[i] = 5;
        laserdata_.intensities[i] = 100;
    }
}

// init scanpub
void LaserScanPub::scanpub_init()
{
    scanpub_ = nh_.advertise<sensor_msgs::LaserScan>("scan",100);
    ros::Rate rate(10);
    while(nh_.ok())
    {
        laserdata_init();
        //publish laser data
        scanpub_.publish(laserdata_);
        rate.sleep();
    }
}

int main(int argc,char**argv)
{
    //init and name node
    ros::init(argc,argv,"myslam_laser");
    //create ros node handle
    ros::NodeHandle nh;
    // 0 -→ 0 degree and 1.57 -→ 90 degree
```

```
    LaserScanPub scanpub(nh,0,1.57,0.01,0,10,100);
    return 0;
}
```

说明如下。

- LaserScanPub::LaserScanPub：构造函数实现激光雷达数据的初始化。
- LaserScanPub::laserdata_init()：填充激光雷达数据。需要注意，前面我们发布坐标变换时将雷达坐标系与机器人坐标系重合。因此，此处的坐标系名称应为“base_link”。
- for(int i=0;i<scanNums;i++)：该循环体填充每一束激光雷达的深度和强度信息。
- while(nh_.ok())：循环发布数据，频率为10Hz。

5.5.4 建图

激光数据及TF坐标变换都具备之后，下面开始编写建图程序。创建book_myslam.h头文件，内容如下：

```
#include "ros/ros.h"
#include "nav_msgs/OccupancyGrid.h"
#include "nav_msgs/MapMetaData.h"
#include "sensor_msgs/LaserScan.h"
#include "tf/transform_listener.h"
#include "tf/tf.h"
#include <vector>
#include <fstream>
#include <math.h>
#include <boost/thread/thread.hpp>
#include <boost/thread/mutex.hpp>

using namespace std;
//define point for obtain point set with bresenham
struct MapPoint
{
    int x,y;
    MapPoint()
    {
        x=0;
        y=0;
    }
```

```
    MapPoint(int x0,int y0)
    {
        x=x0;
        y=y0;
    }
};

// define MySlam class
class MySlam
{
    public:
        MySlam(ros::NodeHandle& nh,double mapreso,double mposx,double mposy,
               double mposz,double morientx,double morienty,double morientz,
               double morientw,int mwidth,int mheight);
        ~MySlam();
        void mappub_init();
        void lasersub_init();
        void lasercallback(const sensor_msgs::LaserScanConstPtr& laserdata);
        void mapdata_init();
        vector<MapPoint> bresenham(int x0,int y0,int x1,int y1);

    private:
        ros::NodeHandle nh_;
        ros::Subscriber lasersub_;
        ros::Publisher mappub_;
        tf::TransformListener tflistener_;
        nav_msgs::OccupancyGrid mapdata_;
        //define map reso, position:x,y,z,orientation:x,y,x,w
        double mapreso,mposx,mposy,mposz,
               morientx,morienty,morientz,
               morientw;
        //define map width height
        int mwidth,mheight;
        vector<MapPoint> endpoints;
        MapPoint endpoint;
        vector<MapPoint> mappoints;
        tf::StampedTransform base2map;
        tf::Quaternion quat;
        double theta;
        tf::Vector3 trans_base2map;
        double tx,ty;
        int basex0,basey0;
```

```
        //scan beams end coordination in laser and map frame;
        double basex,basey;
        double mapx,mapy;
        //laser beams angle in laser frame
        double beamsAngle;
        int mapxn,mapyn;
        int laserNum;
        int nx,ny;
        int idx;
        //save data to file as log for problem check
        ofstream fopen;
        int scan_count;
        int scan_reso;
        boost::mutex map_mutex;
};
```

说明如下。

定义 MapPoint 结构体用于描述地图中的点。

创建 MySlam 类，用于控制建图过程。

- void mappub_init()：初始化地图发布者。
- void lasersub_init()：创建激光雷达接收数据。
- void lasercallback(const sensor_msgs::LaserScanConstPtr& laserdata)：激光雷达数据接收处理回调函数。
- void mapdata_init()：地图数据初始化。
- vector<MapPoint> bresenham(int x0,int y0,int x1,int y1)：利用 bresenham 算法存储激光雷达扫描过的地图栅格点。

下面的 private 成员变量主要定义了地图尺寸、雷达数据点等基本信息。

创建 book_myslam.cpp 文件，完成头文件中所定义方法的具体实现，代码内容如下：

```
#include "book_myslam.h"

MySlam::MySlam(ros::NodeHandle& nh,double mapreso,double mposx,double mposy,
               double mposz,double morientx,double morienty,double morientz,
               double morientw,int mwidth,int mheight):nh_(nh),mapreso(mapreso),
               mposx(mposx),mposy(mposy),mposz(mposz),morientx(morientx),
               morienty(morienty),morientz(morientz),morientw(morientw),
               mwidth(mwidth),mheight(mheight)
{
    mapdata_init();
```

```
        mappub_init();
        lasersub_init();

    }

    MySlam::~MySlam()
    {
    }

    //laser data cllback
    void MySlam::lasercallback(const sensor_msgs::LaserScanConstPtr& laserdata)
    {
        /*lookupTransform (const std::string &target_frame, const std::string &source_frame,
                              const ros::Time &time, StampedTransform &transform)
          get transfrom map --- target frame , base_link --- source frame
        */
        if(scan_count %scan_reso == 0)
        {
            try{
                tflistener_.waitForTransform("map","base_link",ros::Time(0),ros::
Duration(3.0));
                tflistener_.lookupTransform("map","base_link",ros::Time(0),base2map);
            }
            catch(tf::TransformException& ex)
            {
                ROS_INFO("%s",ex.what());
                ros::Duration(1.0).sleep();
            }
            boost::mutex::scoped_lock map_lock(map_mutex);

            //get angle in x direction
            quat = base2map.getRotation();
            theta = quat.getAngle();
            //get transform distance in x,y direction
            trans_base2map = base2map.getOrigin();
            tx = trans_base2map.getX();
            ty = trans_base2map.getY();
            //laser origin coordination in map frame
            basex0 = int(tx/mapreso);
            basey0 = int(ty/mapreso);
            laserNum = laserdata->ranges.size();
            fopen.open("data.txt",ios::app);//save points data for problem check
```

```
            if(fopen.is_open())
            {
                cout<<"open file successsful!"<<endl;
            }else
            {
                cout<<"open file fail"<<endl;
            }
            for(int i=0;i<laserNum;i++)
            {
                beamsAngle = laserdata->angle_min+i*laserdata->angle_increment;
                //end point coordination in base from
                basex = laserdata->ranges[i]*cos(beamsAngle);
                basey = laserdata->ranges[i]*sin(beamsAngle);
                //end point coordination in map frame
                mapx = basex*cos(theta) + basey*sin(theta) + tx;
                mapy = basey*cos(theta) - basex*sin(theta) + ty;
                //points grid coordination
                nx = int(mapx/mapreso);
                ny = int(mapy/mapreso);
                mapxn = nx+1;//
                mapyn = ny+1;//
                endpoint.x = mapxn;
                endpoint.y = mapyn;

                fopen<<endpoint.x<<" "<<endpoint.y<<"\n";
                endpoints.push_back(endpoint);
            }
            fopen.close();

            for(vector<MapPoint>::iterator iter = endpoints.begin();iter!=endpoints.
end();iter++)
            {
                mappoints = MySlam::bresenham(basex0,basey0,(*iter).x,(*iter).y);
                cout<<"scan numbers are: "<<endpoints.size()<<endl;
                cout<<"bresenham point nums are:"<<mappoints.size()<<endl;
                cout<<"x0,yo is "<<basex0<<" "<<basey0<<endl;
                cout<<"angle is "<<theta<<endl;
                for(vector<MapPoint>::iterator iter1 = mappoints.begin();iter1!=mappoints.
end();iter1++)
                {
                    idx = mwidth*(*iter1).y+(*iter1).x;
                    cout<<"idx is"<<(*iter1).x<<" "<<(*iter1).y<<endl;
                    mapdata_.data[idx] = 0;
```

```
            }
            mappoints.clear();
        }
        endpoints.clear();
        mappub_.publish(mapdata_);
    }
    scan_count++;
}

// bresenham algorithm obtain int points in line
vector<MapPoint> MySlam::bresenham(int x0,int y0,int x1,int y1)
{
    vector<MapPoint> pp;
    MapPoint p;
    int dx, dy, h, a, b, x, y, flag, t;
    dx = abs(x1 - x0);
    dy = abs(y1 - y0);
    if (x1 > x0) a = 1; else a = -1;
    if (y1 > y0) b = 1; else b = -1;
    x = x0;
    y = y0;
    if (dx >= dy) {
        //0< |k| <=1
        flag = 0;
    } else {
        //|k|>1,exchange dx, dy
        t = dx;
        dx = dy;
        dy = t;
        flag = 1;
    }
    h = 2 * dy - dx;
    for (int i = 1; i <= dx; ++i) {
        p.x = x,p.y=y;
        //cout<<"x,y is "<<x<<" "<<y<<endl;
        pp.push_back(p);
        if (h >= 0) {
            if (flag == 0) y = y + b;
            else x = x + a;
            h = h - 2 * dx;
        }
        if (flag == 0) x = x + a;
        else y = y + b;
```

```
        h = h + 2 * dy;

    }
    return pp;
}

// map publisher init
void MySlam::mappub_init()
{
    mappub_ = nh_.advertise<nav_msgs::OccupancyGrid>("map",100);
}

// laser data subscriber init
void MySlam::lasersub_init()
{
    lasersub_ = nh_.subscribe("scan",1,&MySlam::lasercallback,this);
}

//map data initialize
void MySlam::mapdata_init()
{
    scan_count = 0;
    scan_reso = 1;
    ros::Time currtime = ros::Time::now();
    mapdata_.header.stamp = currtime;
    mapdata_.header.frame_id = "map";
    mapdata_.info.resolution = mapreso;
    mapdata_.info.width = mwidth;
    mapdata_.info.height = mheight;
    mapdata_.info.origin.position.x = mposx;
    mapdata_.info.origin.position.y = mposy;
    mapdata_.info.origin.position.z = mposz;
    mapdata_.info.origin.orientation.x = morientx;
    mapdata_.info.origin.orientation.y = morienty;
    mapdata_.info.origin.orientation.z = morientz;
    mapdata_.info.origin.orientation.w = morientw;
    int dataszie = mwidth*mheight;
    mapdata_.data.resize(dataszie);
    for(int i=0;i<dataszie;i++)
    {
        mapdata_.data[i]=-1;
    }
}
```

```
int main(int argc,char** argv)
{
    //init and name node
    ros::init(argc,argv,"MySlam");
    //create ros node handle
    ros::NodeHandle nh;
    //map param
    double mapreso=0.05,mposx=0,mposy=0,mposz=0,morientx=0,
           morienty=0,morientz=0,morientw=1;
    int mwidth=300,mheight=300;
    //Myslam instantiate object
    MySlam myslam(nh,mapreso,mposx,mposy,mposz,morientx,morienty,
                  morientz,morientw,mwidth,mheight);
    ros::spin();
    return 0;
}
```

说明如下。

定义构造函数用于地图发布、激光雷达数据接收等。

- MySlam::lasercallback()：激光雷达数据接收及处理回调函数。
- if(scan_count %scan_reso == 0)：确定激光雷达数据的接收频率。
- tflistener_.lookupTransform("map","base_link",ros::Time(0),base2map)：获取 map 和 base_link 的坐标变换关系。

获取激光雷达坐标系在 map 中的坐标：

```
basex0 = int(tx/mapreso);
basey0 = int(ty/mapreso);
```

获取每一束激光末端在 map 中的坐标（double 型）：

```
mapx = basex*cos(theta) + basey*sin(theta) + tx;
mapy = basey*cos(theta) - basex*sin(theta) + ty;
```

获取 gridmap 的坐标（int 型）：

```
nx = int(mapx/mapreso);
ny = int(mapy/mapreso);
```

将雷达数据写入 data.txt 文件：

```
fopen<<endpoint.x<<" "<<endpoint.y<<"\n";
```

保存所有的激光雷达末端坐标到 endpoints：

```
endpoints.push_back(endpoint);
```

通过 bresenham 方法获取每束激光穿过的栅格坐标：

```
mappoints = MySlam::bresenham(basex0,basey0,(*iter).x,(*iter).y);
```

将穿过的栅格在地图中的状态填充为 0：

```
mapdata_.data[idx] = 0;
```

Bresenham 算法画线获取激光束穿过的栅格：

```
vector<MapPoint> MySlam::bresenham(int x0,int y0,int x1,int y1)
```

地图数据初始化：

```
void MySlam::mapdata_init()
```

至此，slam 功能包的所有代码已经准备完毕。创建 CMakeLists.txt 文件进行编译，文件内容如下：

```
cmake_minimum_required(VERSION 2.8.3)
project(book_myslam)

find_package(catkin REQUIRED COMPONENTS roscpp rospy std_msgs sensor_msgs nav_
msgs tf geometry_msgs)

catkin_package(
#  INCLUDE_DIRS include
#  LIBRARIES book_myslam
#  CATKIN_DEPENDS other_catkin_pkg
#  DEPENDS system_lib
)

include_directories(
  include ${catkin_INCLUDE_DIRS}
# include
# ${catkin_INCLUDE_DIRS}
)

add_executable(book_myslam
  src/book_myslam.cpp
)
add_dependencies(book_myslam ${${PROJECT_NAME}_EXPORTED_TARGETS} ${catkin_
EXPORTED_TARGETS})
target_link_libraries(book_myslam
```

```
    ${catkin_LIBRARIES}
  )

  add_executable(slam_laser
    src/slam_laser.cpp
  )
  add_dependencies(slam_laser ${${PROJECT_NAME}_EXPORTED_TARGETS} ${catkin_
EXPORTED_TARGETS})
  target_link_libraries(slam_laser
    ${catkin_LIBRARIES}
  )

  add_executable(slam_tfpub
    src/slam_tfpub.cpp
  )
  add_dependencies(slam_tfpub ${${PROJECT_NAME}_EXPORTED_TARGETS} ${catkin_
EXPORTED_TARGETS})
  target_link_libraries(slam_tfpub
    ${catkin_LIBRARIES}
  )
```

为了方便调试，我们可以将编译完成的 3 个可执行程序放进一个 lunch 文件中一起启动。创建 myslam.launch 文件内容如下：

```
<launch>
    <node pkg="book_myslam" type="slam_tfpub" name="tf_pub"/>
    <node pkg="book_myslam" type="slam_laser" name="laser_pub"/>
    <node pkg="book_myslam" type="book_myslam" name="myslam"/>
</launch>
```

5.5.5 建图测试

下面开始测试 slam 功能包，测试步骤如下。

（1）roscore

（2）rostopic pub -r 10 /cmd_vel geometry_msgs/Twist '{linear: {x: 0.003, y: 0, z: 0}, angular: {x: 0, y: 0, z: 0}}'

（3）roslaunch book_myslam myslam.launch

（4）rosrun rviz rviz

选择 map、scan 等 topic，并可视化 TF 坐标树，建图效果如图 5-14 所示。

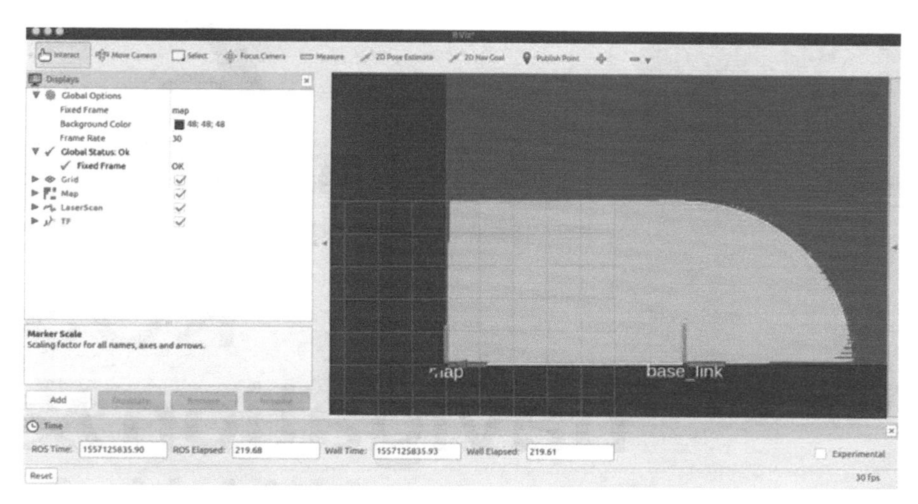

图 5-14　自定义 slam 功能包建图效果

查看节点关系图，如图 5-15 所示。

```
rosrun rqt_graph rqt_graph
```

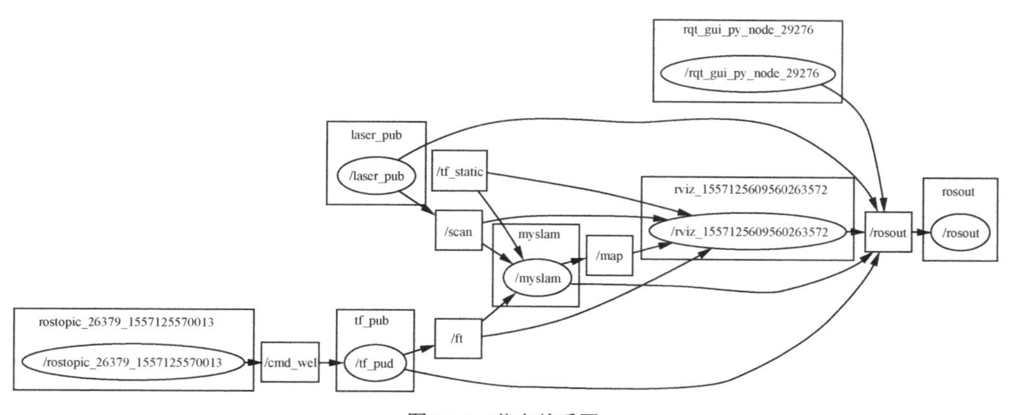

图 5-15　节点关系图

从图 5-15 中可以看出，建图效果非常好。然而，现实中会存在激光测量误差、车轮打滑造成里程计不准的情况，从而导致车体位置坐标不准，建图效果往往不理想。因此，slam 问题的研究重点也多在精确定位等方向。

第6章

ROS navigation及算法简介

6.1 ROS navigation stack概述

在了解了建图的过程及原理之后，本章我们讨论如何利用建好的地图进行导航。ROS 中的导航功能主要集中在 navigation 包中，读者可到 GitHub 中下载该包的源码进行阅读，下载地址为 https://github.com/ros-planning/navigation。

该源码包中包含的功能包及含义如下。

- amcl：根据机器人自身的里程数值以及地图特征，利用粒子滤波修正机器人在已知的地图内的位置。
- base_local_planner：局部路径规划器。
- dwa_local_planner：也是局部路径规划器，使用动态窗口法。
- carrot_planner：很简单的全局路径规划器，生成的路径为目标点到机器人当前点的连线上的点。
- clear_costmap_recovery：无法规划路径的恢复算法。
- costmap_2d：代价地图实现。
- fake_localization：主要用来做定位仿真。
- global_planner：全局路径规划算法包。
- map_server：提供代价地图的管理服务。
- move_base：机器人移动导航框架（导航最主要的逻辑框架）。
- move_slow_and_clear：也是一种恢复策略。
- nav_core：提供接口，能够实现插件更换算法的主要包。
- nav_fn：全局路径规划算法。
- robot_pose_ekf：综合里程计、GPS、imu 数据，通过拓展卡尔曼滤波进行位置估计。
- rotate_recovery：旋转恢复策略实现包。
- voxel_grid：三维代价地图。

整个功能包的原理框架如图 6-1 所示。

在图 6-1 中，白色部分是功能包提供的，灰色模块是可选择的模块，浅蓝色部分是需要用户提供的数据。核心模块包含全局路径规划与局部路径规划，从外部接收的有 map_server 提供的 map 信息、amcl 模块提供基于自适应蒙特卡洛方法的定位功能、sensor 提供激光雷达及视觉等传感器信息、TF 提供里程计及各坐标系之间的变换关系。输出结果是规划出的速度及角速度。

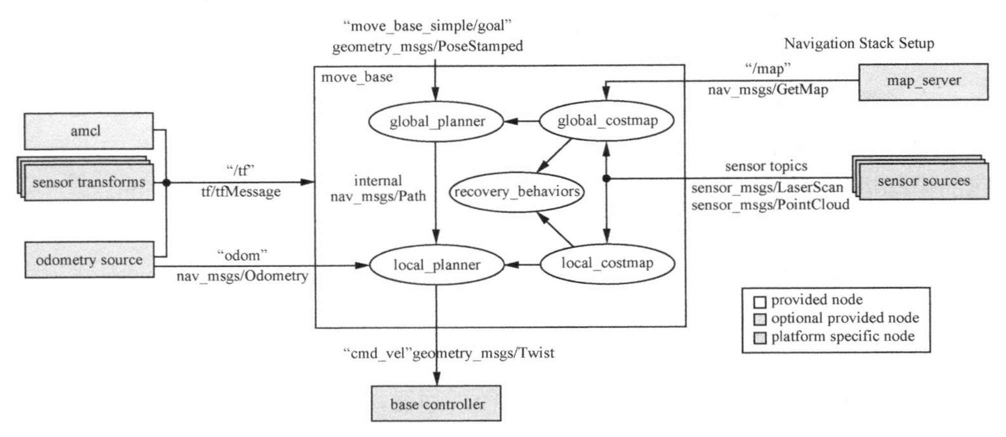

图 6-1 navigation 架构

在 move_base 中，有多种路径规划器算法可选，我们需要告诉 move_base 路径规划器使用哪种算法。全局路径的规划插件如下。

- navfn：ROS 中比较旧的代码，实现了 dijkstra 和 A* 全局规划算法。
- global_planner：重新实现了 Dijkstra 和 A* 全局规划算法，可以看作 navfn 的改进版。
- parrot_planner：一种简单的算法实现全局路径规划算法。

局部路径的规划插件如下。

- base_local_planner：实现了 Trajectory Rollout 和 DWA 两种局部规划算法。
- dwa_local_planner：实现了 DWA 局部规划算法，可以看作 base_local_planner 的改进版本。

6.2 move_base的配置

使用 move_base 功能需要配置 move_base 功能包。首先，创建 Launch 启动文件 move_base.lunch 文件内容如下：

```
<launch>
    <!--run a map server-->
```

```
        <param name="use_sim_time" value="false" />
         <node pkg="map_server" type="map_server" name="map_server_node" args="$(find
myrobot_2dnav_test)/navigation_map/btmap2.pgm 0.05)" />

        <!-- run amcl-->
        <include file="$(find amcl)/examples/amcl_diff.launch"/>

        <!-- config move_base-->
         <node pkg="move_base" type="move_base" name="move_base_node" respawn="false"
output="screen">
                <rosparam file="$(find myrobot_2dnav_test)/map_planner_config/costmap_
common_params.yaml" command="load" ns="global_costmap"/>
                <rosparam file="$(find myrobot_2dnav_test)/map_planner_config/costmap_
common_params.yaml" command="load" ns="local_costmap"/>
                <rosparam file="$(find myrobot_2dnav_test)/map_planner_config/local_
costmap_params.yaml" command="load" />
                <rosparam file="$(find myrobot_2dnav_test)/map_planner_config/global_
costmap_params.yaml" command="load" />
                <!--rosparam file="$(find myrobot_2dnav_test)/map_planner_config/base_
local_planner_params.yaml" command="load" /-->
                <rosparam file="$(find myrobot_2dnav_test)/map_planner_config/dwa_local_
planner_params.yaml" command="load" />
                <rosparam file="$(find myrobot_2dnav_test)/map_planner_config/move_base_
params.yaml" command="load" />
                <rosparam file="$(find myrobot_2dnav_test)/map_planner_config/global_
planner_params.yaml" command="load" />
        </node>
    </launch>
```

启动 map_server 功能节点，提供加载地图功能。

然后，加载 amcl 模块提供定位功能。

> **注意**
>
> amcl 支持两种平台机器人，其中“amcl_omni.launch”对应全向机器人平台，amcl_diff.launch 对应差分驱动机器人平台。

对于这里的 amcl，我们使用的是默认参数配置，若需要自定义参数，也可以按照下面的例子进行相应修改：

```
<launch>
<node pkg="amcl" type="amcl" name="amcl" output="screen">
<param name="odom_model_type" value="diff"/>
<param name="odom_alpha5" value="0.1"/>
```

```
<param name="transform_tolerance" value="0.2" />
<param name="gui_publish_rate" value="10.0"/>
<param name="laser_max_beams" value="30"/>
<param name="min_particles" value="500"/>
<param name="max_particles" value="5000"/>
<param name="kld_err" value="0.05"/>
<param name="kld_z" value="0.99"/>
<param name="odom_alpha1" value="0.2"/>
<param name="odom_alpha2" value="0.2"/>
<!-- translation std dev, m -->
<param name="odom_alpha3" value="0.8"/>
<param name="odom_alpha4" value="0.2"/>
<param name="laser_z_hit" value="0.5"/>
<param name="laser_z_short" value="0.05"/>
<param name="laser_z_max" value="0.05"/>
<param name="laser_z_rand" value="0.5"/>
<param name="laser_sigma_hit" value="0.2"/>
<param name="laser_lambda_short" value="0.1"/>
<param name="laser_lambda_short" value="0.1"/>
<param name="laser_model_type" value="likelihood_field"/>
<!-- <param name="laser_model_type" value="beam"/> -->
<param name="laser_likelihood_max_dist" value="2.0"/>
<param name="update_min_d" value="0.2"/>
<param name="update_min_a" value="0.5"/>
<param name="odom_frame_id" value="odom"/>
<param name="resample_interval" value="1"/>
<param name="transform_tolerance" value="0.1"/>
<param name="recovery_alpha_slow" value="0.0"/>
<param name="recovery_alpha_fast" value="0.0"/>
</node>
</launch>
```

说明如下。

- min_particles 和 max_particles：设定算法运行所允许的粒子的最小和最大数量，如果粒子数多，计算会更加精确，当然也会消耗更多的 CPU 资源。
- laser_model_type：配置激光雷达类型，也可以设置 beam 光束雷达。
- laser_likelihood_max_dist：设置地图中障碍物膨胀的最大距离。

将 amcl 配置完成之后，下面开始配置 move_base 功能模块。首先加载 move_base 功能包，然后加载 costmap_common_params.yaml 文件，将其分别命名为框架中对应的 global_costmap 和 local_costmap。该文件中的内容如下：

```
    map_type: costmap  # add voxel
    max_obstacle_height: 2.0  # add
    min_obstacle_height: 0.0  # add
    #subscribe_to_updates: true  # add
    footprint: [[-0.17,-0.17],[0.17,-0.17],[0.17,0.17],[-0.17,0.17]]
    #robot_radius: ir_of_robot
    #inflation_radius: 0.3
    #obstacle_range: 2.5
    #raytrace_range: 3.0
    #observation_sources: laser_scan_sensor
    #laser_scan_sensor: {sensor_frame: laser, data_type: LaserScan, topic: scane,
marking: true, clearing: true}

    # added newly
    obstacle_layer:
      enabled:              true
      max_obstacle_height:  1.8
      origin_z:             0.0
      z_resolution:         0.0
      z_voxels:             2
      unknown_threshold:    15
      mark_threshold:       0
      combination_method:   1
      track_unknown_space:  true    #true needed for disabling global path planning
through unknown space
      obstacle_range: 2.5
      raytrace_range: 3.0
      origin_z: 0.0
      z_resolution: 0.2
      z_voxels: 2
      publish_voxel_map: true  # default false
      observation_sources:  scan
      scan:
        sensor_frame: laser
        data_type: LaserScan
        topic: scan_filtered
        marking: true
        clearing: true
        #inf_is_valid: true  # add
        min_obstacle_height: 0.0
        max_obstacle_height: 1.8

    inflation_layer:
```

```
    enabled:              true  # true
    cost_scaling_factor:  10.0  # 5.0 exponential rate at which the obstacle cost drops off (default: 10)
    inflation_radius:     0.5 #0.5 max. distance from an obstacle at which costs are incurred for planning paths.

  static_layer:
    enabled:              true  #true
```

该文件主要设置了全局地图和局部地图中的一些公共参数，包含障碍物的高度和宽度、机器人底盘形状（四角点坐标或半径）。obstacle_layer 用于设置了障碍物属性及激光雷达信息，inflation_layer 用于设置膨胀半径。

接下来，lunch 文件加载了全局地图的属性文件 global_costmap_params.yaml，该文件配置内容如下：

```
global_costmap:
  global_frame: /map
  robot_base_frame: base_link
  update_frequency: 3.0
  publish_frequency: 0
  rolling_window: false #  false
  transform_tolerance: 0.5 # 1.0
  static_map: true
  origin_x: 0
  origin_y: 0
  plugins:
   - {name: static_layer,        type: "costmap_2d::StaticLayer"}
   - {name: obstacle_layer,      type: "costmap_2d::VoxelLayer"}
   - {name: inflation_layer,     type: "costmap_2d::InflationLayer"}
```

其中，主要设置了全局坐标系、机器人基坐标系、地图更新频率、地图原点等信息。

配置完全局地图之后是局部地图配置，配置文件为 local_costmap_params.yaml，文件内容如下：

```
local_costmap:
  global_frame: /odom
  robot_base_frame: base_link
  update_frequency: 3.0   #3.0
  publish_frequency: 3.0  # 2.0
  static_map: false
  rolling_window: true
```

```
  width: 2.0
  height: 2.0
  resolution: 0.02
  transform_tolerance: 0.5
  plugins:
   - {name: obstacle_layer,      type: "costmap_2d::VoxelLayer"}
   - {name: inflation_layer,     type: "costmap_2d::InflationLayer"}
```

其中，设置了局部地图的长宽、分辨率及坐标系名称等。

接下来开始设置dwa局部路径规划器配置文件dwa_local_planner_params.yaml，文件内容如下：

```
DWAPlannerROS:

# Robot Configuration Parameters - Kobuki
  max_vel_x: 0.30
  min_vel_x: 0.01

  max_vel_y: 0
  min_vel_y: 0

  max_trans_vel: 0.30 # choose slightly less than the base's capability
  min_trans_vel: 0.01  # this is the min trans velocity when there is negligible
rotational velocity
  trans_stopped_vel: 0.01

  # Warning!
  #   do not set min_trans_vel to 0.0 otherwise dwa will always think translational
velocities
  #   are non-negligible and small in place rotational velocities will be
created.

  max_rot_vel: 1.0 # choose slightly less than the base's capability 0.6
  min_rot_vel: 0.1 # this is the min angular velocity when there is negligible
translational velocity 0.4
  rot_stopped_vel: 0.1

  acc_lim_x: 1 # maximum is theoretically 2.0, but we
  acc_lim_theta: 1.5
  acc_lim_y: 0      # diff drive robot

# Goal Tolerance Parameters
  yaw_goal_tolerance: 0.2
```

```
      xy_goal_tolerance: 0.15
      latch_xy_goal_tolerance: false

    # Forward Simulation Parameters
      sim_time: 2.0       # 1.7
      vx_samples: 10       # 3
      vy_samples: 1       # diff drive robot, there is only one sample
      vtheta_samples: 20  # 20

    # Trajectory Scoring Parameters
      path_distance_bias: 64.0      # 32.0   - weighting for how much it should
stick to the global path plan
      goal_distance_bias: 24.0     # 24.0   - wighting for how much it should attempt
to reach its goal
      occdist_scale: 0.4                # 0.01   - weighting for how much the
controller should avoid obstacles
      forward_point_distance: 0.325 # 0.325  - how far along to place an additional
scoring point
      stop_time_buffer: 0.2         # 0.2    - amount of time a robot must stop in
before colliding for a valid traj.
      scaling_speed: 0.25           # 0.25   - absolute velocity at which to start
scaling the robot's footprint
      max_scaling_factor: 0.2         # 0.2     - how much to scale the robot's
footprint when at speed.

    # Oscillation Prevention Parameters
      oscillation_reset_dist: 0.05  # 0.05   - how far to travel before resetting
oscillation flags

    # Debugging
      publish_traj_pc : true
      publish_cost_grid_pc: true
      global_frame_id: odom

    # Differential-drive robot configuration - necessary?
      holonomic_robot: false
```

该文件主要设置了dwa算法中的最大最小速度、加速度、角速度、采样数、角度及位置误差阈值等信息。

全局路径规划配置文件为global_planner_params.yaml，文件内容如下：

```
  GlobalPlanner:          # Also see: http://wiki.ros.org/global_planner
    old_navfn_behavior: false                         # Exactly mirror behavior of
navfn, use defaults for other boolean parameters, default false
```

```
    use_quadratic: true                              # Use the quadratic approximation
of the potential. Otherwise, use a simpler calculation, default true
    use_dijkstra: true                                   # Use dijkstra's algorithm.
Otherwise, A*, default true
    use_grid_path: false                              # Create a path that follows the
grid boundaries. Otherwise, use a gradient descent method, default false

    allow_unknown: false                               # Allow planner to plan through
unknown space, default true
                                                        #Needs to have track_unknown_
space: true in the obstacle / voxel layer (in costmap_commons_param) to work
    planner_window_x: 0.0                              # default 0.0
    planner_window_y: 0.0                              # default 0.0
    default_tolerance: 0.5                             # If goal in obstacle, plan to
the closest point in radius default_tolerance, default 0.0

    publish_scale: 100                                 # Scale by which the published
potential gets multiplied, default 100
    planner_costmap_publish_frequency: 0.0             # default 0.0

    lethal_cost: 253                                   # default 253
    neutral_cost: 50 #66                                   # default 50
    cost_factor: 3.0 #0.55                                   # Factor to multiply each
cost from costmap by, default 3.0
    publish_potential: true                              # Publish Potential Costmap
(this is not like the navfn pointcloud2 potential), default true
```

其中，主要设置了全局路径规划算法 A* 或 Dijkstra，以及规划更新频率等信息。

最后使用 move_base_params.yaml 文件明确指出所使用的全局及局部路径规划算法，文件内容如下：

```
shutdown_costmaps: false

controller_frequency: 5.0
controller_patience: 3.0

planner_frequency: 1.0
planner_patience: 5.0

oscillation_timeout: 10.0
oscillation_distance: 0.2
```

```
    # local planner - default is trajectory rollout
    #base_local_planner: "base_local_planner/TrajectoryPlannerROS"
    base_local_planner: "dwa_local_planner/DWAPlannerROS"
    base_global_planner: navfn/NavfnROS  #alternatives: carrot_planner/
CarrotPlanner, global_planner/GlobalPlanner
```

至此，配置文件已经介绍完毕。请读者熟悉这些配置，在下一章中，我们将使用这些配置文件对 V-rep 环境下的小车进行导航测试。接下来，我们对 navigation 源码包进行简单解读。

6.3 navigation源码解读

整个 move_base 的调用入口在源码包的 move_base/src 文件夹下的 move_base_node.cpp 中，如下：

```
ros::init(argc, argv, "move_base_node");
tf2_ros::Buffer buffer(ros::Duration(10));
tf2_ros::TransformListener tf(buffer);

move_base::MoveBase move_base( buffer );
```

接下来，转入同目录下的 move_base.cpp 文件，打开后，首先完成参数的初始化：

```
    MoveBase::MoveBase(tf2_ros::Buffer& tf) :
      tf_(tf),
      as_(NULL),
      planner_costmap_ros_(NULL), controller_costmap_ros_(NULL),
      bgp_loader_("nav_core", "nav_core::BaseGlobalPlanner"),
      blp_loader_("nav_core", "nav_core::BaseLocalPlanner"),
      recovery_loader_("nav_core", "nav_core::RecoveryBehavior"),
      planner_plan_(NULL), latest_plan_(NULL), controller_plan_(NULL),
      runPlanner_(false), setup_(false), p_freq_change_(false), c_freq_change_
(false), new_global_plan_(false)
```

接下来，获取目标点函数：

```
void MoveBase::goalCB(const geometry_msgs::PoseStamped::ConstPtr& goal)
```

获取机器人位置坐标函数：

```
MoveBase::getRobotPose(geometry_msgs::PoseStamped& global_pose, costmap_2d::Costmap2DROS* costmap)
```

从起始点到目标点的全局路径规划在下面的makePlan函数中执行：

```
bool MoveBase::makePlan(const geometry_msgs::PoseStamped& goal, std::vector<geometry_msgs::PoseStamped>& plan)
```

在该函数中执行全局路径规划：

```
if(!planner_->makePlan(start, goal, plan) || plan.empty())
```

执行全局路径规划之后，转入局部路径规划，进入executeCycle函数，该函数原型如下：

```
bool MoveBase::executeCycle(geometry_msgs::PoseStamped& goal, std::vector<geometry_msgs::PoseStamped>& global_plan)
```

在该函数中，将全局路径传入局部路径规划器中，如果传入失败，将出现报错提示：

```
if(!tc_->setPlan(*controller_plan_)){
  //ABORT and SHUTDOWN COSTMAPS
  ROS_ERROR("Failed to pass global plan to the controller, aborting.");
  resetState();

  //disable the planner thread
  lock.lock();
  runPlanner_ = false;
  lock.unlock();

  as_->setAborted(move_base_msgs::MoveBaseResult(), "Failed to pass global plan to the controller.");
  return true;
}
```

tc_ 为局部路径规划器，定义在move_base.h中：

```
boost::shared_ptr<nav_core::BaseLocalPlanner> tc_;
```

然后，在switch(state_)中根据状态PLANNING:、CONTROLLING:、CLEARING:分别执行路径规划、控制及状态恢复。局部路径规划入口如下：

```
if(tc_->computeVelocityCommands(cmd_vel)){
    ROS_DEBUG_NAMED( "move_base", "Got a valid command from the local planner: %.3lf, %.3lf, %.3lf",cmd_vel.linear.x, cmd_vel.linear.y, cmd_vel.angular.z );
```

```
        last_valid_control_ = ros::Time::now();
        //make sure that we send the velocity command to the base
        vel_pub_.publish(cmd_vel);
```

下面进入 dwa 局部路径规划。打开 dwa_local_planner/src 文件夹下的 dwa_planner_ros.cpp 文件，找到 computeVelocityCommands 函数：

```
    bool DWAPlannerROS::computeVelocityCommands(geometry_msgs::Twist& cmd_vel)
```

首先，通过局部路径规划器获取全局路径在局部地图下的映射，映射的末端即为局部子目标点。映射失败时会有报错提示：

```
    if ( ! planner_util_.getLocalPlan(current_pose_, transformed_plan)) {
        ROS_ERROR("Could not get local plan");
        return false;
      }
```

然后，更新局部路径及局部地图：

```
    dp_->updatePlanAndLocalCosts(current_pose_,transformed_plan,costmap_ros_-
>getRobotFootprint());
```

最后，通过 dwa 算法规划速度：

```
    bool isOk = dwaComputeVelocityCommands(current_pose_, cmd_vel);
```

该函数原型如下：

```
    bool   DWAPlannerROS::dwaComputeVelocityCommands(geometry_msgs::PoseStamped &global_
pose, geometry_msgs::Twist& cmd_vel)
```

在该函数中，通过评价函数获取最优路径及速度：

```
    base_local_planner::Trajectory path = dp_->findBestPath(global_pose, robot_
vel, drive_cmds);
```

dp_ 的定义在 dwa_planner_ros.h 文件中，表示 dwa 路径规划器：

```
    boost::shared_ptr<DWAPlanner> dp_;
```

接下来，打开 dwa_planner.cpp 文件找到 findBestPath 函数：

```
    base_local_planner::Trajectory DWAPlanner::findBestPath(
          const geometry_msgs::PoseStamped& global_pose,
          const geometry_msgs::PoseStamped& global_vel,
          geometry_msgs::PoseStamped& drive_velocities) {
```

首先，产生速度空间：

```
// prepare cost functions and generators for this run
generator_.initialise(pos,
    vel,
    goal,
    &limits,
    vsamples_);
```

generator_ 对应的定义如下：

```
base_local_planner::SimpleTrajectoryGenerator generator_;
```

然后找到base_local_planner/src文件下的simple_trajectory_generator.cpp文件，即可找到该函数产生速度空间的原理。

最后，通过打分器对速度空间中每个速度对应的路径进行打分，获取最优路径及对应的速度：

```
scored_sampling_planner_.findBestTrajectory(result_traj_, &all_explored);
```

scored_sampling_planner_ 定义如下：

```
base_local_planner::SimpleScoredSamplingPlanner scored_sampling_planner_;
```

打开base_local_planner/src文件夹下的simple_scored_sampling_planner.cpp文件，找到findBestTrajectory对应的实现内容：

```
bool  SimpleScoredSamplingPlanner::findBestTrajectory(Trajectory& traj, std::vector
<Trajectory>* all_explored)
```

该函数中gen_list_ 和critics_ 对应的定义如下：

```
std::vector<TrajectorySampleGenerator*> gen_list_;
std::vector<TrajectoryCostFunction*> critics_;
```

TrajectorySampleGenerator 和 TrajectoryCostFunction 的定义分别在base_local_planner/include/base_local_planner/文件夹下的trajectory_sample_generator.h和trajectory_ cost_function.h中，它们的功能分别是根据速度产生规划轨迹、对轨迹进行打分。当打开这两个文件时，你可能会觉得困惑，因为这两个文件中除了类的基本定义之外，并没有定义如何产生轨迹以及如何打分，这是因为它们都是父类的定义，这时打开同目录下的simple_trajectory_generator.h，你会看到如下内容：

```
class SimpleTrajectoryGenerator: public base_local_planner::TrajectorySample
Generator {
```

可以看到 SimpleTrajectoryGenerator 继承于 TrajectorySampleGenerator。现在打开 /base_local_planner/src 文件夹下的 simple_trajectory_generator.cpp 文件：

```
bool SimpleTrajectoryGenerator::hasMoreTrajectories()
bool SimpleTrajectoryGenerator::nextTrajectory(Trajectory &comp_traj)
bool SimpleTrajectoryGenerator::generateTrajectory(
      Eigen::Vector3f pos,
      Eigen::Vector3f vel,
      Eigen::Vector3f sample_target_vel,
      base_local_planner::Trajectory& traj) {
```

在 simple_scored_sampling_planner.cpp 文件中：

```
TrajectorySampleGenerator* gen_ = *loop_gen;
      while (gen_->hasMoreTrajectories()) {
        gen_success = gen_->nextTrajectory(loop_traj);
```

这段代码的引用来自于 simple_trajectory_generator.cpp。找到 generateTrajectory，你会看到根据速度生成规划路径的代码：

```
pos = computeNewPositions(pos, loop_vel, dt);
```

该函数的实现原理（实际就是航迹推演）如下：

```
Eigen::Vector3f SimpleTrajectoryGenerator::computeNewPositions(const Eigen::
Vector3f& pos,const Eigen::Vector3f& vel, double dt) {
     Eigen::Vector3f new_pos = Eigen::Vector3f::Zero();
     new_pos[0] = pos[0] + (vel[0] * cos(pos[2]) + vel[1] * cos(M_PI_2 +
pos[2])) * dt;
     new_pos[1] = pos[1] + (vel[0] * sin(pos[2]) + vel[1] * sin(M_PI_2 +
pos[2])) * dt;
     new_pos[2] = pos[2] + vel[2] * dt;
     return new_pos;
}
```

再看 base_local_planner/include/base_local_planner 目录下的如下文件。

- map_grid_cost_function.h
- obstacle_cost_function.h
- oscillation_cost_function.h
- twirling_cost_function.h

打开它们之后，你会发现它们都继承自 TrajectoryCostFunction：

```
class MapGridCostFunction: public base_local_planner::TrajectoryCostFunction {
class ObstacleCostFunction : public TrajectoryCostFunction {
```

```
class OscillationCostFunction: public base_local_planner::TrajectoryCostFunction {
class TwirlingCostFunction: public base_local_planner::TrajectoryCostFunction {
```

准备好打分项之后，就可以分别调用 scoreTrajectory 函数来筛选最优路径了。

```
loop_traj_cost = scoreTrajectory(loop_traj, best_traj_cost);
```

该函数原型如下：

```
double SimpleScoredSamplingPlanner::scoreTrajectory(Trajectory& traj, double best_
traj_cost) {
```

首先，在 for 循环遍历各个打分项：

```
for(std::vector<TrajectoryCostFunction*>::iterator score_function = critics_.
begin(); score_function != critics_.end(); ++score_function) {
    TrajectoryCostFunction* score_function_p = *score_function;
```

然后，利用每个打分项的 scoreTrajectory 函数对轨迹进行打分。如果读者对各个打分项的打分原理感兴趣，那么可以到对应的文件中阅读，这里不再详述。

```
double cost = score_function_p->scoreTrajectory(traj);
```

将每个得分项乘以比例系数并相加，得到总分：

```
if (cost != 0) {
  cost *= score_function_p->getScale();
}
traj_cost += cost;
```

找到最优路径对应的得分：

```
        if (traj_cost > best_traj_cost) {
          break;
return traj_cost;
```

最后，将最优路径对应的速度发布出去即可。至此，我们已经完成了 move_base 中路径规划源码的全部解读。接下来，将介绍全局路径规划算法 A-Star、局部路径规划算法 dwa 的原理及其具体实现。

6.4 A-Star算法原理与实现

A-Star（A*）算法是一种求解最短路径最有效的直接搜索方法，也是许多其他问题的常用启发式算法。该算法综合了最良优先搜索和 Dijkstra 算法的优点，在进行启发式搜索提高算法效率的同时，可以保证找到一条最优路径（基于评估函数）。A* 算法核心公式就是 F 值的计算：

$$F = G + H$$

- F：方块的总移动代价（从初始状态经由状态 n 到目标状态的代价估计）。
- G：开始点到当前方块的移动代价（状态空间中从初始状态到状态 n 的实际代价）。
- H：当前方块到结束点的预估移动代价（从状态 n 到目标状态的最佳路径的估计代价）。

该公式具有以下特性。

- 如果 G 为 0，即只计算任意顶点 n 到目标的评估函数 H，而不计算起点到顶点 n 的距离，则算法转化为使用贪心策略的最良优先搜索，速度最快，但可能得不到最优解。
- 如果 H 不大于顶点 n 到目标顶点的实际距离，则一定可以求出最优解，而且 H 越小，需要计算的节点越多，算法效率越低。常见的评估函数有欧几里得距离、曼哈顿距离、切比雪夫距离。
- 如果 H 为 0，即只需求出起点到任意顶点 n 的最短路径 G，而不计算任何评估函数 H，则转化为单源最短路径问题，即 Dijkstra 算法，此时需要计算的顶点最多。

A* 算法的伪码如下。

（1）将开始点记录为当前点 P。

（2）将当前点 P 放入封闭列表。

（3）搜寻点 P 的所有邻近点，假如某邻近点没有在开放列表或封闭列表中，则计算出该邻近点的 F 值，并设父节点为 P，然后将其放入开放列表。

（4）判断开放列表是否已经空了，如果没有，说明在达到结束点前已经找完了所有可能的路径点，则寻路失败，算法结束；否则，继续。

（5）从开放列表拿出一个 F 值最小的点，作为寻路路径的下一步。

（6）判断该点是否为结束点，如果是，则寻路成功，算法结束；否则，继续。

（7）将该点设为当前点 P，跳回到第 3 步。

> **注意**
>
> 在第 3 步中，如果邻近点在开放列表中，则需要检查这条路径（即经由当前方格到达它那里）是否更好，用 G 值作参考。更小的 G 值表示这是更好的路径，如果是这样，则把它的父亲设置为当前方格，并重新计算它的 G 值和 F 值，然后更新 G 值和 F 值。

该算法的流程框图如图 6-2 所示。其中，O 代表开放列表，C 代表封闭列表。

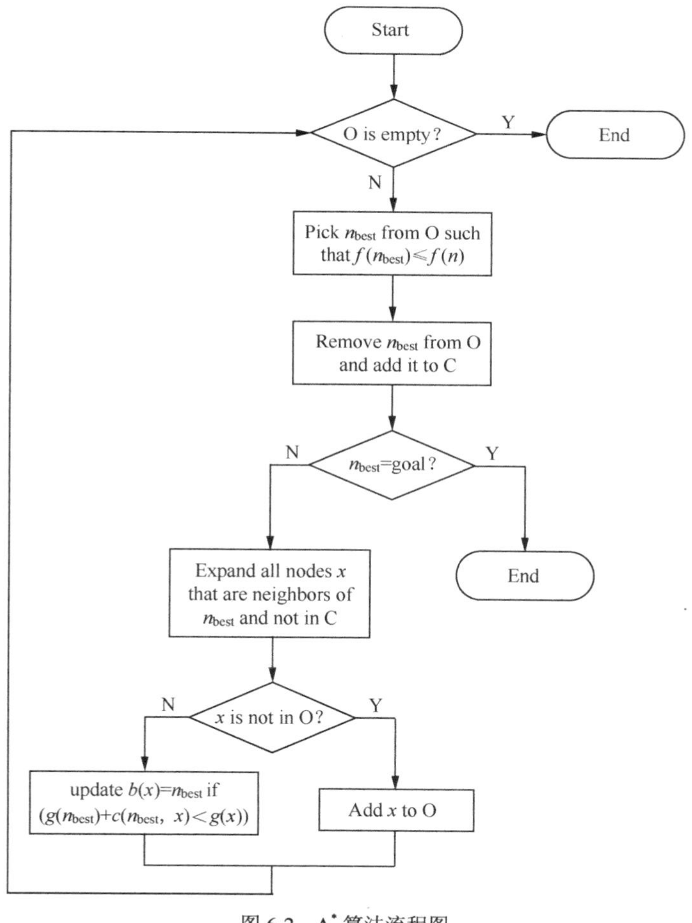

图 6-2 A* 算法流程图

接下来，我们通过一个简单的例子来阐述 A* 算法的实现过程。

在图 6-3 中，A 点为起点，N 点为终点。图中各点到 N 点的 H 值在右上角列出。线上两点间连线上的数值代表两点的实际代价 G。

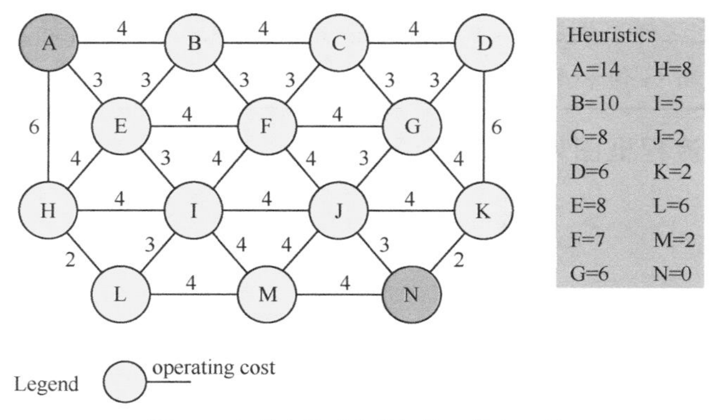

图 6-3 起始点及对应的初始 H 值和 G 值

图 6-4 表示将 A 点加入封闭列表，并更新相邻点的值。

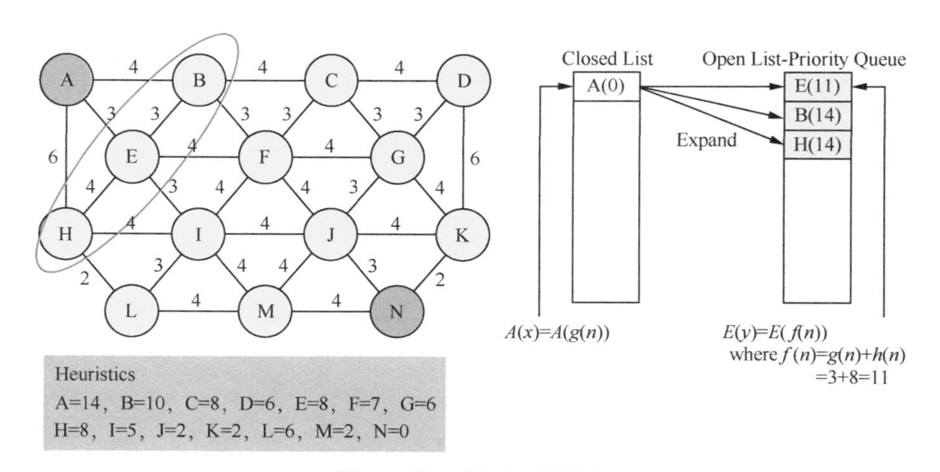

图 6-4 将 A 点加入封闭列表

将 A 点放入封闭列表后，计算 A 点所有相邻点放入开放列表的 F 值，将开放列表中 F 值最小的 E 点移除并添加进封闭列表，如图 6-5 所示。

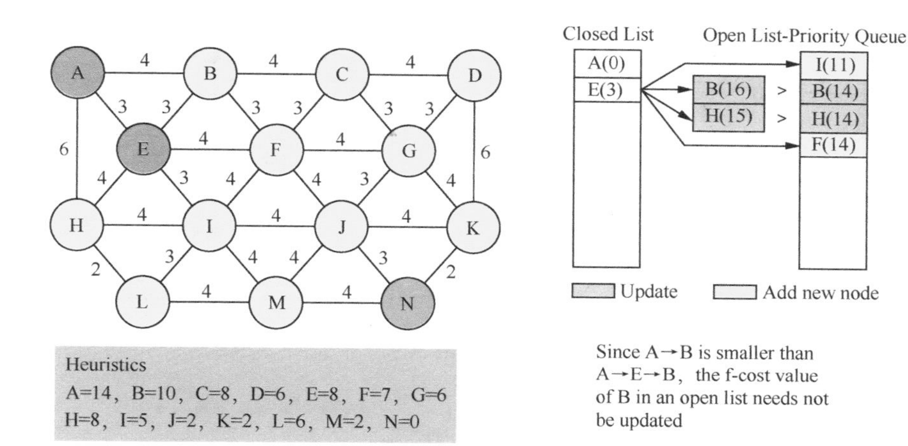

图 6-5　将 E 点加入封闭列表

计算 E 的所有相邻点 F 值，结合相邻点是否在开放列表及新的 G 值大小，更新开放列表中的点及对应的 G 值。此时，相邻点中 I 点的 F 值最小，将其从开放列表移除并放入封闭列表，如图 6-6 所示。

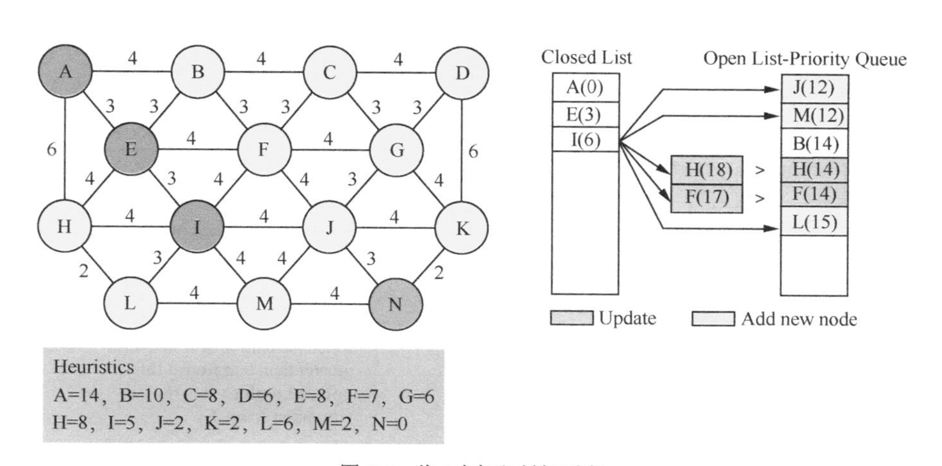

图 6-6　将 I 点加入封闭列表

同理，得到 J 点的 F 值最小，将其从开放列表移除并添加到封闭列表，如图 6-7 所示。

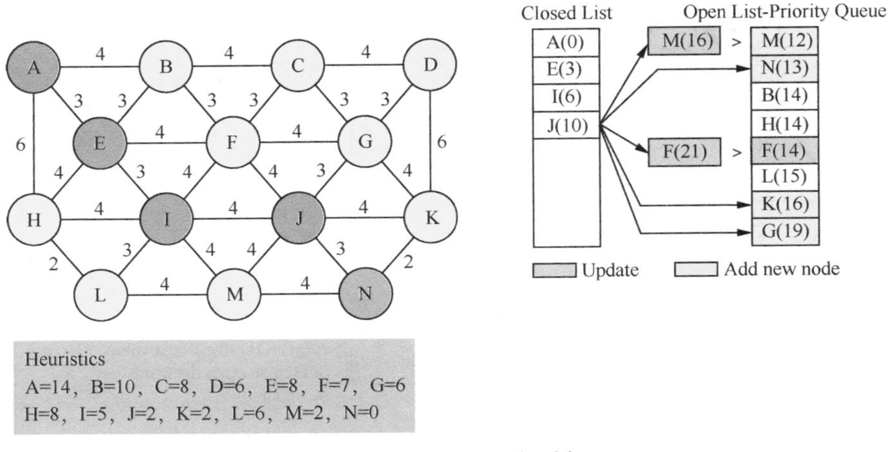

图 6-7 将 J 点加入封闭列表

最后，得到目标点 N，如图 6-8 所示。

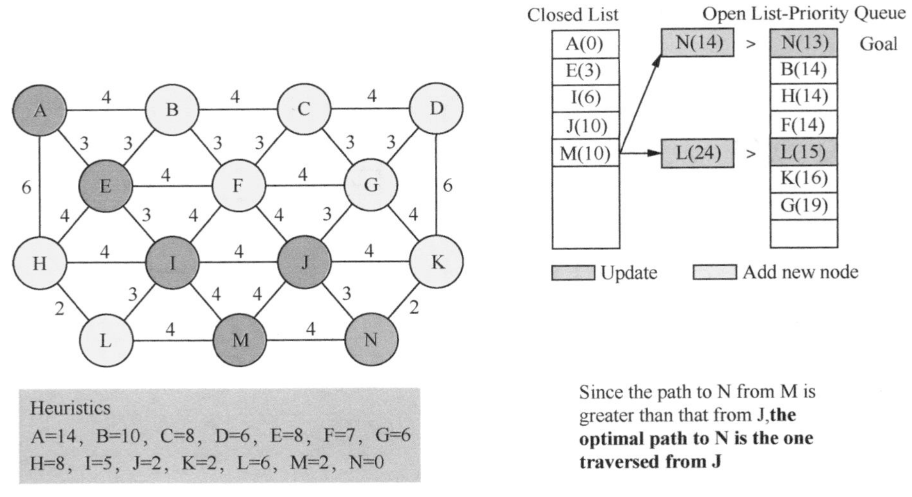

图 6-8 将目标点 N 加入封闭列表

接下来，我们利用 Python 语言来实现该算法，并利用 matplot 模块将搜集过程实时可视化。新建 astar.py 文件，内容如下：

```
import matplotlib.pyplot as plt
import math

#define node
class Node:
 def __init__(self,x,y,parent,cost,index):
        self.x = x
        self.y = y
        #used for tracking path node in closeset
        self.parent = parent
        self.cost = cost
        # node index
        self.index = index

# tracking path node
def calc_path(goaln,closeset):
rx,ry=[goaln.x],[goaln.y]
print closeset[-1]
#goal parent node in closeset
parentn = closeset[-1]

while parentn!=None:
        rx.append(parentn.x)
        ry.append(parentn.y)
        parentn = parentn.parent
return rx,ry


# astar algorithm core
def astar_plan(sx,sy,gx,gy):
ox,oy,xwidth,ywidth=map_generation()
# show map
plt.figure('Astar algorithm demo')
plt.plot(ox,oy,'ks')
plt.plot(sx,sy,'bs')
plt.plot(gx,gy,'ro')
# define motion model
motion = motion_model()
# define open and close set for saving node
openset,closeset=dict(),dict()
sidx = sy*xwidth+sx
gidx = gy*xwidth+gx
# start node and goal node
```

```
    starn=Node(sx,sy,None,0,sidx)
    goaln=Node(gx,gy,None,0,gidx)
    openset[sidx] = starn
    while 1:
            # calculate node with min f_cost as current node
            c_id = min(openset,key=lambda o:openset[o].cost + h_cost(openset[o],goaln))
            curnode = openset[c_id]
            # if arrive goal point break else closset add current node and remove
current node from openset
            if curnode.x == goaln.x and curnode.y == goaln.y:
                print 'find goal'
                closeset[-1] = curnode
                break
            else:
                closeset[c_id] = curnode
                plt.plot(curnode.x,curnode.y,'gx')
                if len(openset.keys())%10==0:
                    plt.pause(0.01)
                del openset[c_id]

            # check 8 direction point
            for j in range(len(motion)):
                newnode = Node(curnode.x+motion[j][0],
                               curnode.y+motion[j][1],
                               curnode,
                               curnode.cost + motion[j][2],
                               c_id)
                n_id = index_calc(newnode,xwidth)

                # if node in closeset out of loop once
                if n_id in closeset:
                    continue

                # if node in obstacle out of loop once
                if node_verify(newnode,ox,oy):
                    continue

                # if node not in openset add it to openset
                #  else compare it with old one and update it with lower g_cost and
update parent also
                if n_id not in openset:
                    openset[n_id] = newnode
                else:
                    if openset[n_id].cost >= newnode.cost:
```

```
                    openset[n_id] = newnode
# get path node x ,y
px,py = calc_path(goaln,closeset)
return px,py

# Map generation
def map_generation():
# ox,oy list for obstacles
ox,oy=[],[]
for i in range(60):
        ox.append(i)
        oy.append(0)

for i in range(60):
        ox.append(i)
        oy.append(60)

for i in range(60):
        ox.append(0)
        oy.append(i)

for i in range(60):
        ox.append(60)
        oy.append(i)

for i in range(25):
        ox.append(i)
        oy.append(20)

for i in range(40):
        ox.append(35)
        oy.append(i)

for i in range(40):
        ox.append(50)
        oy.append(60-i)

minx = min(ox)
miny = min(oy)
maxx = max(ox)
maxy = max(oy)
# map  xwidth,  ywidth
xwidth = maxx-minx
ywidth = maxy-miny
```

```
return ox,oy,xwidth,ywidth



# motion model in 8 direction with 8 g_costs
def motion_model():
motion =[[1,0,1],
             [1,1,math.sqrt(2)],
             [1,-1,math.sqrt(2)],
             [0,1,1],
             [0,-1,1],
             [-1,1,math.sqrt(2)],
             [-1,0,1],
             [-1,-1,math.sqrt(2)]]
return motion

def h_cost(node,goal):
# Weight
w = 1.0
# Euclidean distance be careful: this distance will cost longer time
#h = w*math.sqrt((goal.x-node.x)**2 + (goal.y-node.y)**2)
# Mahattan distance
h = w*(abs(goal.x-node.x) + abs(goal.y-node.y))
return h

# Node index obtain
def index_calc(node,xwid):
n_id = node.y*xwid + node.x
return n_id

# Verify node is in obstacle or not
def node_verify(node,ox,oy):
if (node.x,node.y) in zip(ox,oy):
     return True
else:
    return False

def main():
# Define  start point(sx,sy),goal point(gx,gy)
sx,sy=15,15
gx,gy=55,50
# Astar algorithm tracking path node
```

```
rx,ry=astar_plan(sx,sy,gx,gy)
print rx,ry
# show path
plt.plot(rx,ry,'r-',linewidth=3)
plt.show()

if __name__ =="__main__":
main()
```

说明如下。

定义 Node 节点类，包含坐标、父节点、代价，以及索引等信息。

- def calc_path(goaln,closeset)：根据每个点的父节点信息，从目标点将所有的路径点坐标找出来。
- def map_generation()：产生地图数据，包含障碍物坐标点。
- def motion_model()：定义运动模型，向每个节点相邻的 8 个方向运动，并生成运动代价。
- def h_cost(node,goal)：定义评价函数，可采用欧氏距离或曼哈顿距离。
- def index_calc(node,xwid)：计算节点的索引。
- def node_verify(node,ox,oy)：确定节点是否为障碍物。
- def astar_plan(sx,sy,gx,gy)：A^* 算法核心，参数为起始点坐标。
- openset,closeset=dict(),dict()：开放列表和封闭列表初始化。
- openset[sidx] = starn：将起点加入开放列表。
- c_id = min(openset,key=lambda o:openset[o].cost + h_cost(openset[o],goaln))：在开放列表中找到代价最小的节点对应的索引。
- curnode = openset[c_id]：将代价最小的节点作为考察节点。
- plt.plot(curnode.x,curnode.y,'gx')：用绿色交叉点画出考察点。
- if curnode.x == goaln.x and curnode.y == goaln.y：是否到达终点。
- closeset[c_id] = curnode：将考察点放入封闭列表（注意以键值对形式存储在字典中）。
- for j in range(len(motion))：考察相邻的 8 个邻近点。
- px,py = calc_path(goaln,closeset)：计算路径上所有点的坐标。
- plt.plot(rx,ry,'r-',linewidth=3)：将路径以红色线条画出。

运行程序，可看到如图 6-9 所示的结果。

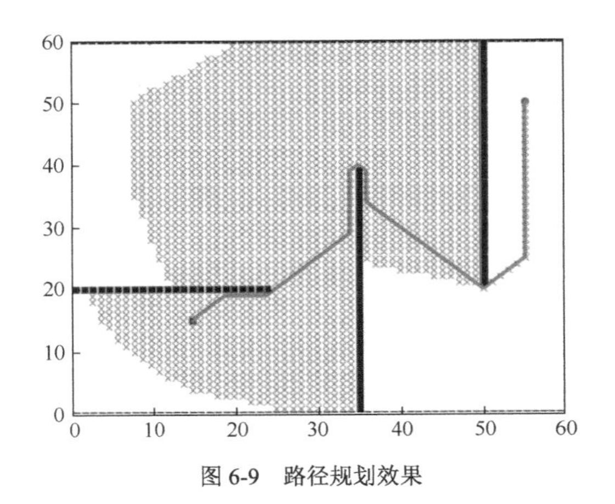

图 6-9 路径规划效果

6.5 dwa算法

机器人局部路劲规划方法有很多，动态窗口法（dynamic window approach，dwa）是最常用的方法之一。该算法主要是在速度（v,w）空间中采样多组速度，并模拟机器人在这些速度下一定时间（sim_period）内的轨迹。在得到多组轨迹后，对这些轨迹进行评价，选取最优轨迹所对应的速度来驱动机器人运动。该算法的突出点在于“动态窗口”这个名词，它的含义是依据移动机器人的加减速性能，限定速度采样空间在一个可行的动态范围内。由以下 3 组评价函数对每条轨迹进行评价。

- 生成轨迹与参考路径的距离（贴合程度）。
- 生成轨迹与参考路径终点的距离。
- 生成轨迹上是否存在障碍物（若有，则抛弃这条轨迹）。

优点如下。

- 反应速度较快，计算不复杂，通过速度组合（线速度与角速度）可以快速得出下一时刻规划轨迹的最优解。
- 可以由横向与纵向两个维度向一个维度优化。

缺点如下。

- 较高的灵活性会极大地降低运动的平稳性。

1. 机器人运动模型

在动态窗口算法中，模拟机器人的轨迹前需要知道机器人的运动模型。这里假设两轮移动机器人的轨迹是一段一段的圆弧或者直线（旋转速度为 0 时），一对（v_t, w_t）就代表一个圆弧轨迹。假设机器人不是全向移动的，即不能纵向移动，只能前进和旋转（v_t, w_t）。在计算机器人轨迹时，先考虑两个相邻时刻，如图 6-10 所示。简单起见，由于机器人相邻时刻（一般码盘采样周期 ms 计）内的运动距离短，因此可以将两个相邻点之间的运动轨迹看成直线，具体推导过程如下。

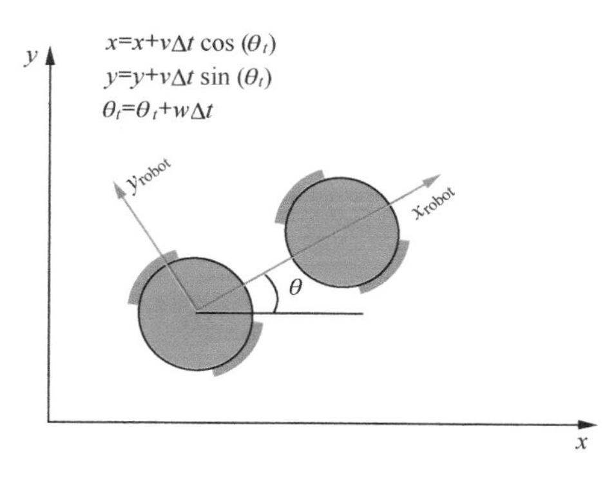

图 6-10　机器人运动模型

如果机器人是全向运动的，即 y 轴有了速度，只需将机器人在机器人坐标 y 轴移动的距离投影到世界坐标系：

$$\begin{cases} \Delta x = v_y \Delta t \cos\left(\theta_t + \dfrac{\pi}{2}\right) = -v_y \Delta t \sin(\theta_t) \\ \Delta y = v_y \Delta t \sin\left(\theta_t + \dfrac{\pi}{2}\right) = -v_y \Delta t \cos(\theta_t) \end{cases}$$

此时的轨迹推演只需将 y 轴移动的距离叠加在之前的计算公式上：

$$\begin{cases} x = x + v\Delta t \cos(\theta_t) - v_y \Delta t \sin(\theta_t) \\ y = y + v\Delta t \sin(\theta_t) + v_y \Delta t \cos(\theta_t) \\ \theta_t = \theta_t + w\Delta t \end{cases}$$

若要得到更精确的计算结果，则应假设相邻时间段内机器人的轨迹是圆弧，而不是直线。圆弧运动的半径为：

$$r = \frac{v}{w}$$

当旋转速度 w 不等于 0 时，机器人的坐标为：

$$
\begin{cases}
x = x - \frac{v}{w}\sin(\theta_t) + \frac{v}{w}\sin(\theta_t + w\Delta t) \\
y = y - \frac{v}{w}\cos(\theta_t) - \frac{v}{w}\cos(\theta_t + w\Delta t) \\
\theta_t = \theta_t + w\Delta t
\end{cases}
$$

2. 速度采样

得到机器人的轨迹运动模型之后，根据速度就可以推算出轨迹。因此只需采样很多速度，推算轨迹，然后评价这些轨迹。

如何采样速度是 dwa 算法的第二个核心：在速度（v,w）的二维空间中存在无穷多组速度，但是根据机器人本身的限制和环境限制，可以将采样速度控制在一定范围内。

移动机器人受自身最大速度和最小速度的限制：

$$V_m = \{v \in [v_{\min}, v_{\max}], w \in [w_{\min}, w_{\max}]\}$$

移动机器人会受到电机性能的影响，由于电机力矩有限，存在最大的加减速限制，因此移动机器人轨迹前向模拟的周期 sim_period 内，存在一个动态窗口，在该窗口内的速度是机器人能够实际达到的速度：

$$V_d = \left\{ (v,w) \middle| \begin{array}{c} v \in [v_c - \dot{v}_b\Delta t, v_c + \dot{v}_a\Delta t] \wedge \\ w \in [w_c - \dot{w}_b\Delta t, w_c + \dot{w}_a\Delta t] \end{array} \right\}$$

其中，v_c 和 w_c 是机器人的当前速度，其他标志对应最大加速度和最大减速度。

为了能够在碰到障碍物前停下来，因此在最大减速度条件下，速度有一个范围：

$$V_a = \{(v,w) \mid v \leqslant \sqrt{2 * dist(v,w) * \dot{v}_b} \wedge w \leqslant \sqrt{2 * dist(v,w) * \dot{w}_b}\}$$

动态窗口采样轨迹如图 6-11 所示。

3. 评价函数

在采样的速度组中，有若干组轨迹是可行的，因此采用评价函数的方式为每条轨迹进行评价。评价函数如下：

$$G(v,w) = \sigma(\alpha * heading(v,w) + \beta * dist(v,w) + \gamma * velocity(v,w))$$

方位角评价函数用来评价机器人在当前设定的采样速度下，达到模拟轨迹末端时的朝向和目标之间的角度差距，如图 6-12 所示。

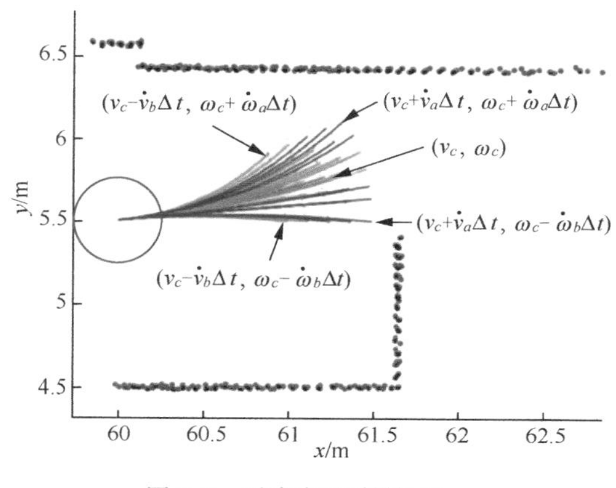

图 6-11 动态窗口采样轨迹

图 6-12 角度差距

空隙 *dist*（*v,w*）代表机器人在当前轨迹上与最近的障碍物之间的距离。如果在这条轨迹上没有障碍物，那么就将其设定一个常数。在实际使用时，如果轨迹上有障碍物，则直接舍弃该路径。

velocity（*v,w*）用来评价当前轨迹的速度大小。我们希望尽可能快速移动到目标点，同时也需要注意速度跳变带来的运动抖动。

上面 3 个部分计算出来后不是直接相加，而是每个部分在归一化后，再相加。机器人传感器检测到的最小障碍物距离在二维空间中是不连续的，这条轨迹能够遇到障碍，旁边那边不一定能遇到。并且这条轨迹最小的障碍物距离是 1m，旁边那条就是 10m。障碍物距离的这种评价标准导致评价函数不连续，也会导致某个项在评价函数中太占优势，比如这里的离障碍物距离 10m 相对于 1m 就太占优势了。所以，要通过将它们归一化来消除某项在评价函数中过分占优。

$$normal\_head(i)=\frac{head(i)}{\sum_{i=1}^{n}head(i)}$$

$$normal\_dist(i)=\frac{dist(i)}{\sum_{i=1}^{n}dist(i)}$$

$$normal\_velocity(i)=\frac{velocity(i)}{\sum_{i=1}^{n}velocity(i)}$$

在介绍完算法原理之后，我们依然用 Python 语言来实现并可视化，模拟 dwa 实际工作效果。新建 dwa.py 文件，文件内容如下：

```
import numpy as np
import matplotlib.pyplot as plt
import math

class Info():
 def __init__(self):
        # define robot move speed ,accelerate,radius ...and so on
        self.v_min = -0.5
        self.v_max = 3.0
        self.w_max = 50.0 * math.pi / 180.0
        self.w_min = -50.0 * math.pi / 180.0
        self.vacc_max = 0.5
        self.wacc_max = 30.0 * math.pi / 180.0
        self.v_reso = 0.01
        self.w_reso = 0.1 * math.pi / 180.0
        self.radius = 1.0
        self.dt = 0.1
        self.predict_time = 4.0
        self.goal_factor = 1.0
        self.vel_factor = 1.0
        self.traj_factor = 1.0

def motion_model(x,u,dt):
# robot motion model: x,y,theta,v,w
x[0] += u[0] * dt * math.cos(x[2])
x[1] += u[0] * dt * math.sin(x[2])
x[2] += u[1] * dt
x[3] = u[0]
x[4] = u[1]

return x

def vw_generate(x,info):
#generate v,w window for traj prediction
Vinfo = [info.v_min,info.v_max,
         info.w_min,info.w_max]

Vmove = [x[3] - info.vacc_max * info.dt,
         x[3] + info.vacc_max * info.dt,
          x[4] - info.wacc_max * info.dt,
```

```
            x[4] + info.wacc_max * info.dt]

    vw = [max(Vinfo[0],Vmove[0]),min(Vinfo[1],Vmove[1]),
          max(Vinfo[2],Vmove[2]),min(Vinfo[3],Vmove[3])]

    return vw

    def traj_cauculate(x,u,info):
    ctraj = np.array(x)
    xnew = np.array(x) # Caution!!! Don't use like this: xnew = x, it will change x
value when run motion_modle below
    time = 0

    while time <= info.predict_time:
          xnew = motion_model(xnew,u,info.dt)
          ctraj = np.vstack((ctraj,xnew))
          time += info.dt

    return ctraj

    def dwa_core(x,u,goal,info,obstacles):
    # the kernel of dwa
    vw = vw_generate(x,info)
    best_ctraj = np.array(x)
    min_score = 10000.0

    for v in np.arange(vw[0],vw[1],info.v_reso):
          for w in np.arange(vw[2],vw[3],info.w_reso):
                # cauculate traj for each given (v,w)
                ctraj = traj_cauculate(x,[v,w],info)
                # cauculate the current traj score
                goal_score = info.goal_factor * goal_evaluate(ctraj,goal)
                vel_score = info.vel_factor * velocity_evaluate(ctraj,info)
                traj_score = info.traj_factor * traj_evaluate(ctraj,obstacles,info)
                ctraj_score = goal_score + vel_score + traj_score
                #evaluate current traj (the score smaller,the traj better)
                if min_score >= ctraj_score:
                    min_score = ctraj_score
                    u = np.array([v,w])
                    best_ctraj = ctraj

    return u,best_ctraj
```

```
def goal_evaluate(traj,goal):
#cauculate current pose to goal with euclidean distance
goal_score = math.sqrt((traj[-1,0]-goal[0])**2 + (traj[-1,1]-goal[1])**2)

return goal_score

def velocity_evaluate(traj,info):
#cauculate current velocty score
vel_score = info.v_max - traj[-1,3]
return vel_score

def traj_evaluate(traj,obstacles,info):
#evaluate current traj with the min distance to obstacles
min_dis = float("Inf")
for i in range(len(traj)):
      for ii in range(len(obstacles)):
            current_dist = math.sqrt((traj[i,0] - obstacles[ii,0])**2 + (traj[i,1]
- obstacles[ii,1])**2)

            if current_dist <= info.radius:
                return float("Inf")

            if min_dis >= current_dist:
                min_dis = current_dist

return 1 / min_dis

def obstacles_generate():
obstacles = np.array([[0,10],
                          [2,10],
                          [4,10],
                          [6,10],
                          [3,5],
                          [4,5],
                          [5,5],
                          [6,5],
                          [7,5],
                          [8,5],
                          [10,7],
                          [10,9],
                          [10,11],
                          [10,13]])
return obstacles
```

```
def local_traj_display(x,goal,current_traj,obstacles):
#display current pose ,traj prodicted,map,goal
plt.cla()
plt.plot(goal[0],goal[1],'or',markersize=10)
plt.plot([0,14],[0,0],'-k',linewidth=7)
plt.plot([0,14],[14,14],'-k',linewidth=7)
plt.plot([0,0],[0,14],'-k',linewidth=7)
plt.plot([14,14],[0,14],'-k',linewidth=7)
plt.plot([0,6],[10,10],'-y',linewidth=10)
plt.plot([3,8],[5,5],'-y',linewidth=10)
plt.plot([10,10],[7,13],'-y',linewidth=10)
plt.plot(obstacles[:,0],obstacles[:,1],'*b',markersize=8)
plt.plot(x[0],x[1],'ob',markersize=10)
plt.arrow(x[0],x[1],math.cos(x[2]),math.sin(x[2]),width=0.02,fc='red')
plt.plot(current_traj[:,0],current_traj[:,1],'-g',linewidth=2)
plt.grid(True)
plt.pause(0.001)

def main():
x = np.array([2,2,45*math.pi/180,0,0])
u = np.array([0,0])
goal = np.array([8,8])
info = Info()
obstacles = obstacles_generate()
global_traj = np.array(x)
plt.figure('DWA Algorithm')
for i in range(2000):
      u,current_traj = dwa_core(x,u,goal,info,obstacles)
      x = motion_model(x,u,info.dt)
      global_traj = np.vstack((global_traj,x))
      local_traj_display(x,goal,current_traj,obstacles)
      if math.sqrt((x[0]-goal[0])**2 + (x[1]-goal[1])**2) <= info.radius:
            print "Goal Arrived!"
            break

plt.plot(global_traj[:,0],global_traj[:,1],'-r')
plt.show()

if __name__ == "__main__":
main()
```

说明如下。

- class Info()：定义Info类对象，说明机器人运动极限速度、加速度等信息。
- def motion_model(x,u,dt)：机器人运动模型。
- def vw_generate(x,info)：速度空间产生原理。
- def traj_cauculate(x,u,info)：根据当前位置及速度预测轨迹。
- def goal_evaluate(traj,goal)：距离目标点评价。
- def velocity_evaluate(traj,info)：速度评价。
- def traj_evaluate(traj,obstacles,info)：轨迹距离障碍物距离评价。
- def obstacles_generate()：地图及障碍物生成。
- def local_traj_display(x,goal,current_traj,obstacles)：局部路径可视化。
- def dwa_core(x,u,goal,info,obstacles)：dwa 算法核心。
- vw = vw_generate(x,info)：产生速度空间。
- ctraj = traj_cauculate(x,[v,w],info)：根据当前位置及速度生成轨迹。
- goal_score = info.goal_factor * goal_evaluate(ctraj,goal)：轨迹末端距离目标点评价。
- vel_score = info.vel_factor * velocity_evaluate(ctraj,info)：当前速度评价。
- traj_score = info.traj_factor * traj_evaluate(ctraj,obstacles,info)：轨迹距离障碍物远近评价。
- ctraj_score = goal_score + vel_score + traj_score：轨迹总分。
- if min_score >= ctraj_score：根据分数筛选最优轨迹。
- for i in range(2000)：设置仿真时间。
- u,current_traj = dwa_core(x,u,goal,info,obstacles)：获取最优轨迹及其对应速度。
- global_traj = np.vstack((global_traj,x))：将机器人当前位置加入全局路径中。
- local_traj_display(x,goal,current_traj,obstacles)：局部路径可视化。
- if math.sqrt((x[0]-goal[0])**2 + (x[1]-goal[1])**2) <= info.radius：判断是否到达目标点。
- plt.plot(global_traj[:,0],global_traj[:,1],'-r')：将机器人经过的点以红色点连接出来，形成实际轨迹。

需要注意，为了简化程序，我们在计算总分时并未使用归一化对各项分数进行处理。读者可对该程序进行修改，对比观察效果。此外，这里我们假设机器人不存在定位误差，但现实运动中会存在轮子打滑等现象，造成机器人存在定位误差，读者可以在此程序基础上对机器人位置进行高斯噪声叠加，再检查该算法的效果。

运行程序，可以看到图 6-13 和图 6-14 所示的效果。

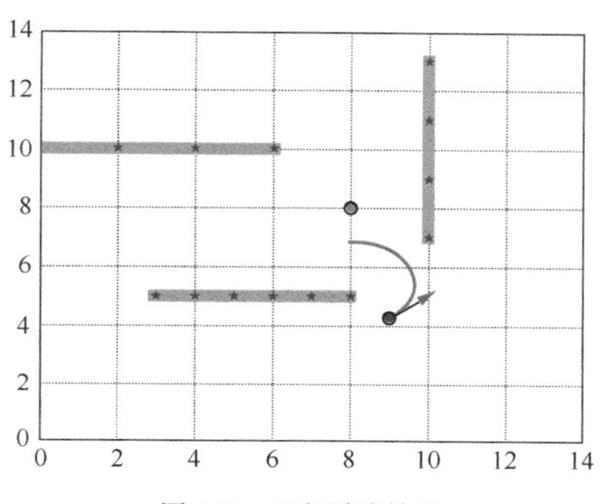

图 6-13　局部避障效果

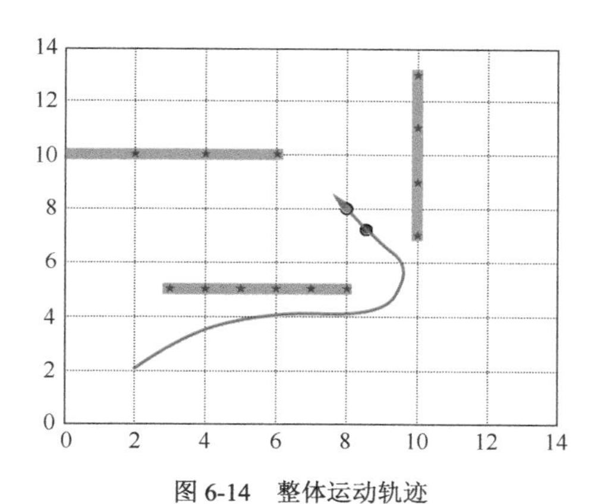

图 6-14　整体运动轨迹

从图中可以看到，绕过障碍物的是绿色规划轨迹，以及从起点到终点机器人所经过的是红色路径，这证明我们的算法可以满足基本的避障要求。

第 7 章

基于 V-rep 的 ROS 开发

7.1 V-rep机器人仿真软件概述

V-rep 是一个强大的机器人三维集成开发环境，号称机器人仿真器里的“瑞士军刀”（软件界面如图 7-1 所示）。它是基于分布式控制架构的、免费的、完善的开发环境，内部集成工业串联机械臂、并联机械臂、多足机器人、移动机器人模型，同时也可以根据用户需要导入对应的 Solidworks 模型。

图 7-1 V-rep 软件界面

目前，国内并无太多有关 V-rep 软件的使用说明或相关教程，与 ROS 结合使用的资料更是少之又少。因此，本书希望通过一些实例操作展示 V-rep 的强大功能，以满足读者对利用 V-rep 开发机器人的需求。V-rep 软件架构如图 7-2 所示。

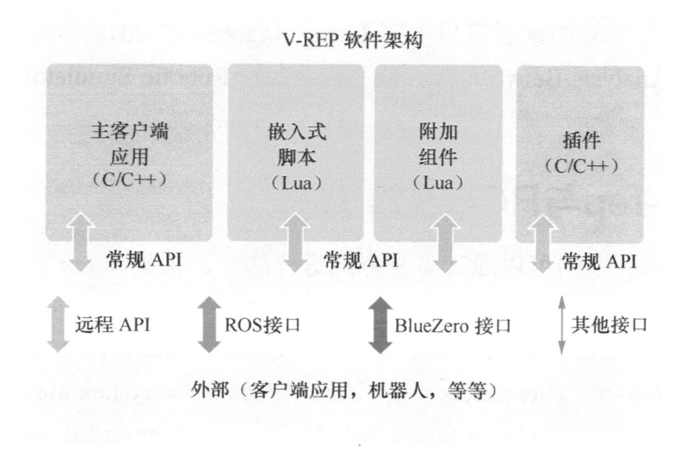

图 7-2 V-rep 软件架构

图 7-2 展示了 V-rep 具有丰富的 API，这些接口或基于 C/C++ 语言，或基于 Lua 脚本。作为快速原型验证、远程监控、快速算法开发、机器人相关教育和工厂自动化系统仿真的工具，V-rep 主要具有以下优点。

- 跨平台（Windows、macOS、Linux）。
- 支持多种编程方法（嵌入式脚本、插件、附加组件、ROS 节点、远程客户端应用编程接口或自定义的解决方案）。
- 多种编程开发语言（C/C++、Python、Java、Lua、Matlab、Octave 和 Urbi）。
- 丰富应用编程接口函数。
- 提供 ROS 服务、发布类型、ROS 订户类型、可拓展性好。
- 完整的运动学解算器（对于任何机构的逆运动学和正运动学）。
- 路径规划。
- 多种图像处理的视觉传感器。
- 数据记录与可视化（时距图、X/Y 图或三维曲线）。

除此之外，还有其他功能特色，这里不作赘述。

7.1.1 V-rep与Gazebo的区别

Gazebo 的优势在于完全兼容 ROS，但是构建机器人仿真环境较复杂。相比于 Gazebo，V-rep 有丰富的场景（各种机器人、传感器、移动平台等）可供使用，这可以使学习者省去大量模型搭建的时间，从而将更多的精力投入算法研究中去。这也是我们选择 V-rep 作为 ROS 机器人仿真软件的重要原因。有关二者的具体不同，有兴趣的读者可以参考 Lucas Nogueira 在 2014 年发表的一篇论文“Comparative Analysis Between Gazebo and V-REP Robotic Simulators”，本书不作详细讨论。

7.1.2 V-rep与ROS通信机制

ROS 和 V-rep 之间可以通过多种方式进行通信。根据 V-rep 官网的介绍，主要有以下 3 种通信方式。

- The RosInterface: the RosInterface duplicates the C/C++ ROS API with a good fidelity. This makes it the ideal choice for very flexible communication via ROS, but might require a little bit more insight on the various messages and the way ROS operates.（官方 The RosInterface 通信）

- The ROS plugin skeleton: this represents a skeleton project that can be used to create a new ROS plugin for V-REP. Make sure to first have a look at the RosInterface source code before attempting to edit this project.（ROS skeleton 插件通信）
- ROS interfaces developed by others: those are not directly supported by us. For instance, the V-REP ROS bridge.（第三方开发的 ROS interfaces 通信）

其中，推荐读者使用第一种（The RosInterface），这也是 V-rep 官方提供并推荐的方式，最后一种（vrep_ros_bridge）是 ros_wiki 上提供的。本书中采用 The RosInterface 方式。

7.2 V-rep安装与ROS配置

7.2.1 环境要求

- 系统版本：Ubuntu14.04（64 位）。
- ROS 版本：indigo。
- V-rep 版本：V-rep 3.4.0 EUD（64 位）。

需要注意，由于 V-rep 与 ROS 的兼容性并不如 Gazebo，所以不同的版本结合效果也不同。这是作者经过尝试以后认为相对稳定的版本组合。此外，目前官网首页所展示的最新的 V-rep 版本是 3.5，如需下载其他版本，可在页面下端找到“Previous versions of V-REP are available here”，单击进入即可下载对应版本。

7.2.2 V-rep的安装

将所下载的安装包解压到合适目录（此处根据读者个人喜好，无特别要求）。

```
~$  cd ~/path_to_your_vrep/
~$  ./vrep.sh
```

运行之后，打开效果如图 7-3 所示。

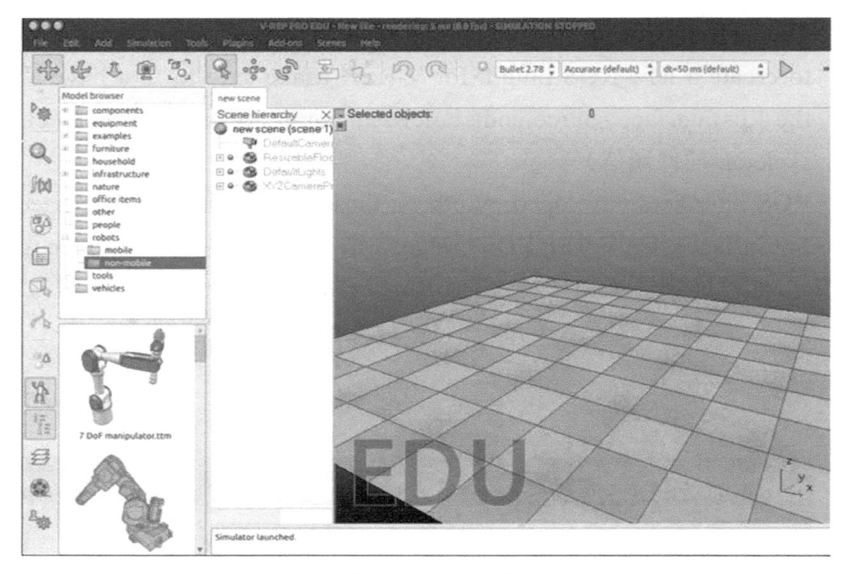

图 7-3 V-rep 界面

这说明你现在已经完成了 V-rep 的初步安装。不过，若要与 ROS 联合使用，我们还需对其进行配置。否则，在终端运行 ./vrep.sh，你将会看到如下加载失败提示：

```
Plugin 'RosInterface': load failed (failed initialization).
Plugin 'RosSkeleton': load failed (failed initialization).
```

接下来，我们将进行相关配置以解决上述问题。

7.2.3 配置RosInterface

转到 V-rep 安装路径下的 compiledRosPlugins 文件夹下：

```
~$  cd ~/path_to_your_vrep/compiledRosPlugins
~$  ls
```

你将看到两个与 ROS 通信相关的文件，分别为 libv_repExtRosInterface.so 和 libv_repExtRosSkeleton.so。上面遇到的加载错误提示就是因为这两个文件没有配置，我们只需稍微改动这两个配置文件的位置即可解决这个问题。

```
cp  ~/path_to_your_vrep/compiledRosPlugins/libv_repExtRosInterface.so  ~/path_
to_your_vrep/
```

```
cp  ~/path_to_your_vrep/compiledRosPlugins/libv_repExtRosSkeleton.so  ~/path_to_your_vrep/
```

或直接将 compiledRosPlugins 文件夹下的所有文件（因为该文件夹下只有我们需要的两个文件，除此之外没有其他文件）直接复制到上一层。

```
~$  cp  ~/path_to_your_vrep/compiledRosPlugins*  ~/path_to_your_vrep/
```

下面测试配置是否成功，打开终端启动 ros master：

```
~$  roscore
```

新打开另一终端，启动 V-rep：

```
~$  cd ~/path_to_your_vrep/
~$     ./vrep.sh
```

如果终端显示如图 7-4 所示的提示信息，看到加载成功提示信息 Plugin 'RosInterface': load succeeded. 和 Plugin 'RosSkeleton': load succeeded，则证明我们已经配置成功了，接下来尝试 V-rep 中提供的 ROS 控制场景。

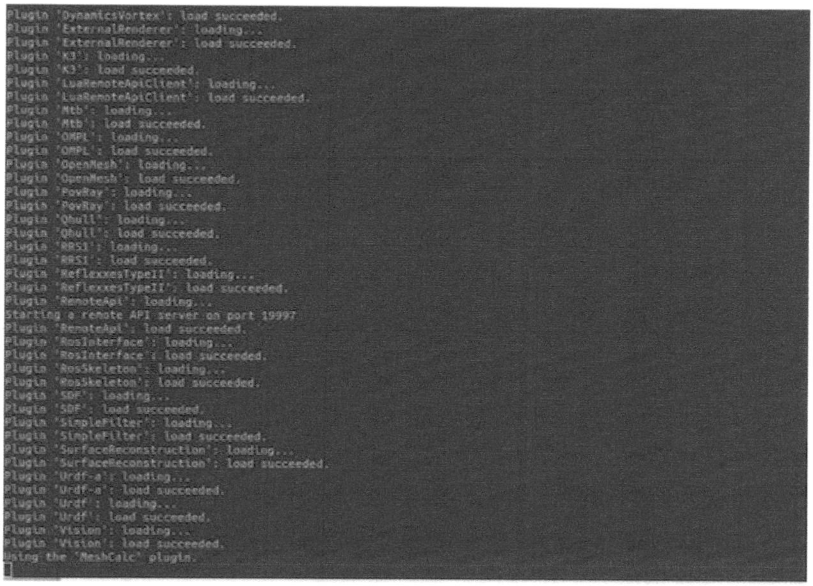

图 7-4　终端提示信息

7.3 运行V-rep自带ROS控制场景

7.3.1 熟悉V-rep基本操作

观察上述 V-rep 软件启动后界面，可以看到最上面有 9 个选项模块，各自所提供的功能如表 7-1 所示。

表 7-1 选项模块及其功能

选项模块	功　能
File	包含场景的打开（最近打开的工程）、关闭、新建以及保存、另存为；模型加载、另存为；导入、导出、退出
Edit	撤销、重做、粘贴、选择、删除、移除
Add	添加几何体（立方体、圆柱、球等）、动力、关节灯光、相机、力传感器；视觉传感器、路径、运动控制脚本等
Simulation	开始、暂停、停止仿真；选择仿真渲染引擎；仿真加速减速及具体设置场景物体、计算模型属性、场景物体操作设置
Tools	脚本、视频监控、用户自定义设置等
Plugins	导入 SDF 和 URDF（ROS 中的机器人模型大多是基于 URDF 文件编写的）
Add-ons	添加功能示例、脚本示例
Scenes	新场景
Help	帮助选项、V-rep 介绍、相关调试

上述选项中，我们常用的是前 3 个，读者掌握基本操作即可，后续会具体展示如何使用这些操作搭建属于我们自己的机器人仿真环境。在选项模块下方提供了一些具体操作选项，如相机的旋转、移动、打开、关闭；场景物体选择、旋转、平移、组合、组合解除；撤销、重做；仿真开始、暂停、加速、减速等。

界面左端的 Model browser 提供了许多 V-rep 自带的场景功能，如组件、设备、传感器、自然环境、办公环境、各种机器人、工具及移动平台。打开相应的 .ttt 场景文件，即可实现场景或模型的加载。

中间的 Scene hierachy 展示了所添加物体的属性，可通过单击实现选择及属性更改。界面右端是场景的可视化展示区，仿真的运动过程也在该区域展示。

7.3.2 运行ROS控制场景

通过上述介绍，相信你已经对 V-rep 有了大致的了解，接下来让我们尝试使用 V-rep 中自带的有关 ROS 控制的场景。打开终端运行 rosore，启动 ros master。重新打开一个终端，并进入你的 V-rep 安装路径，通过 ./vrep.sh 打开软件，通过 .File ->Open scene 打开 rosInterfaceTopicPublisherAndSubscriber.ttt 场景文件，并在界面上方单击 star simulation 的三角形按钮。你将在软件展示区看到如图 7-5 所示的场景。

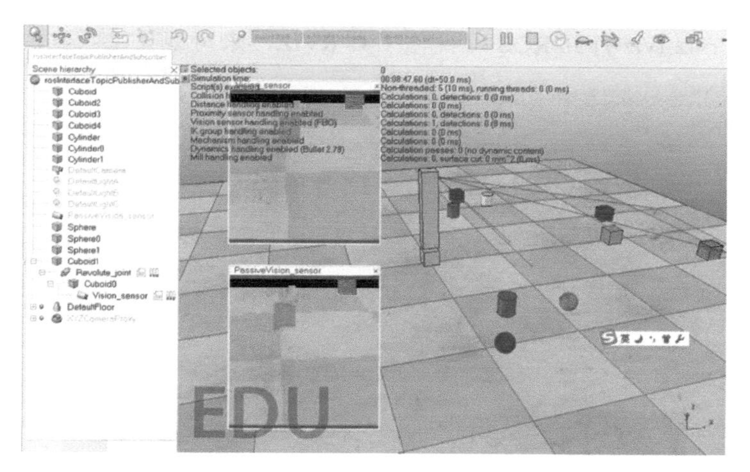

图 7-5　V-rep 中自带 ROS 控制的场景

该 V-rep 场景主要演示了如何仿真一个摄像，并将接收到的信息发布图像话题到 ROS。下面我们查看以下节点之间的通信关系，新开终端运行 rosrun rqt_graph rqt_graph，如图 7-6 所示。

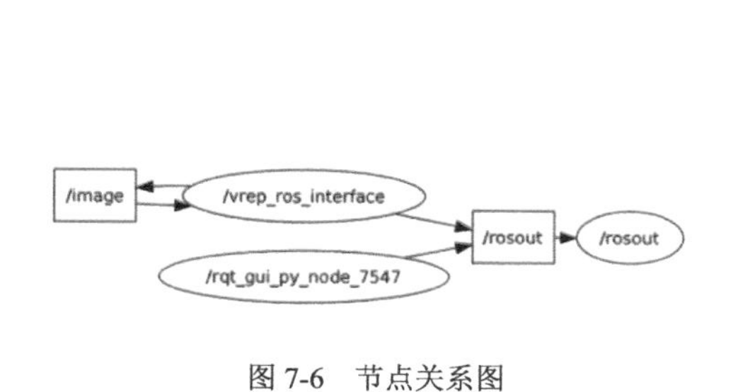

图 7-6　节点关系图

可以看到，/vrep_ros_interface 节点发布了 /imag 的 topic。新建终端运行 rostopic list，查看所有的 topic 如下：

```
/image
/rosout
/rosout_agg
/statistics
/tf
/vrep_ros_interface/addStatusbarMessage
/vrep_ros_interface/objectCount
```

新开终端运行 rostopic echo /image，将该 topic 中的信息打印出来，你也可以运行 rosrun rviz rviz 打开 ROS 可视化工具 RVIZ，来可视化接收到的图像信息，如图 7-7 所示。

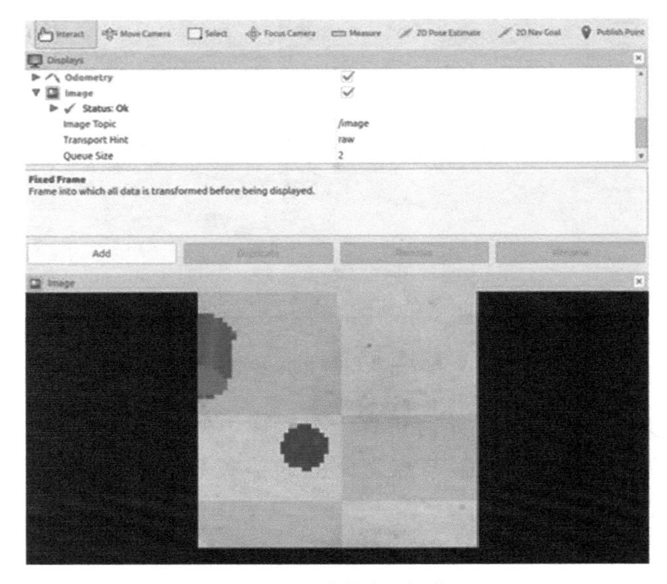

图 7-7 图像信息可视化

有的读者可能已经注意到了，Rviz 中的图像与 V-rep 中的相反。这是数据流的问题造成的，解决这个问题可以从两个方面入手。

（1）在 V-rep 发布端对数据进行置反处理。

（2）在 ROS 创建一个接收节点负责置反接收到的图像信息，并重新发布 topic。

无论怎样，我们已经可以将数据发送给 ROS。接下来，我们将尝试通过 ROS 发布数据到 V-rep。

7.3.3 ROS发送数据到V-rep

同样，单击“File → Open scene”，打开 controlTypeExamples.ttt 场景文件，如图 7-8 所示。

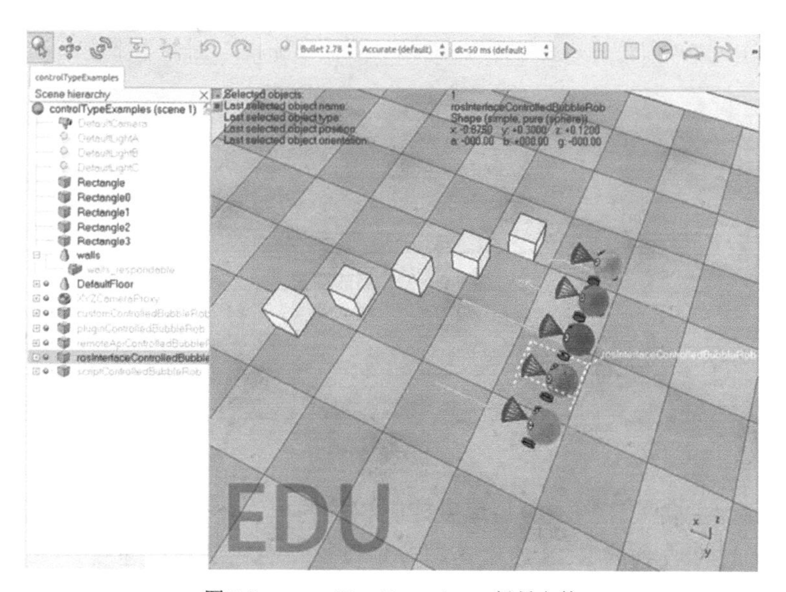

图 7-8 controlTypeExamples.ttt 场景文件

这里展示了 5 种不同的控制方式，有远程 API、脚本控制、rosInterface 接口等。我们主要关注 rosInterfaceControlledBubbleRob，为了防止其他控制方式的干扰，可以删除其他 4 个小车（单击小车选中，按 <Delete> 键即可删除），只保留我们将要控制的 rosInterfaceControlledBubbleRob 小车。

> **注意**
>
> 为了不破坏 V-rep 自带的场景文件，我们不保存做完删除操作的场景。当然，读者也可以复制一份场景文件，在 V-rep 中操作该复制场景，这样更安全。

删除后的场景如图 7-9 所示。

此时，单击仿真按钮，你将发现红色小车在场景中随机运动。为了实现 ROS 端发送速度控制，我们需要对其脚本文件进行修改。双击打开控制脚本，即 rosInterfaceControlledBubbleRob 后面的第一个文本标志，如图 7-10 所示。

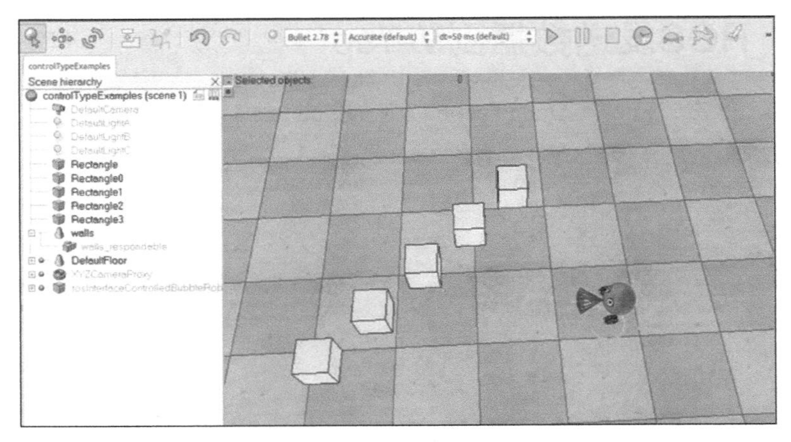

图 7-9　删除后的场景

图 7-10　控制脚本

这里我们主要关注脚本中与速度控制相关的模块，分别是以下部分。

（1）通过接收到的 ROS 速度话题分别设置左右轮的马达速度。

```
function setLeftMotorVelocity_cb(msg)
    -- Left motor speed subscriber callback
    simSetJointTargetVelocity(leftMotor,msg.data)
end

function setRightMotorVelocity_cb(msg)
    -- Right motor speed subscriber callback
    simSetJointTargetVelocity(rightMotor,msg.data)
End
```

（2）负责发送 TF 话题到 ROS。

```
function getTransformStamped(objHandle,name,relTo,relToName)
t=simGetSystemTime()
p=simGetObjectPosition(objHandle,relTo)
o=simGetObjectQuaternion(objHandle,relTo)
return {
      header={
           stamp=t,
           frame_id=relToName
```

```
            },
            child_frame_id=name,
            transform={
                translation={x=p[1],y=p[2],z=p[3]},
                rotation={x=o[1],y=o[2],z=o[3],w=o[4]}
            }
        }
    End
```

（3）分别设置小车、左右轮马达及传感器的控制句柄。

```
    robotHandle=simGetObjectAssociatedWithScript(sim_handle_self)
    leftMotor=simGetObjectHandle("rosInterfaceControlledBubbleRobLeftMotor")
-- Handle of the left motor
    rightMotor=simGetObjectHandle("rosInterfaceControlledBubbleRobRightMotor")
-- Handle of the right motor
    noseSensor=simGetObjectHandle("rosInterfaceControlledBubbleRobSensingNose")
-- Handle of the sensor
```

（4）订阅速度话题或发布传感器和时间话题。

```
    sensorPub=simExtRosInterface_advertise('/'..sensorTopicName,'std_msgs/Bool')
    simTimePub=simExtRosInterface_advertise('/'..simulationTimeTopicName,'std_
msgs/Float32')
                leftMotorSub=simExtRosInterface_subscribe('/'..leftMotorTopicName,'std_
msgs/Float32','setLeftMotorVelocity_cb')
                rightMotorSub=simExtRosInterface_subscribe('/'..rightMotorTopicName,
'std_msgs/Float32','setRightMotorVelocity_cb')
```

这里我们主要修改以下内容。

（1）将以下内容：

```
    local leftMotorTopicName='leftMotorSpeed'..sysTime -- we add a random component
so that we can have several instances of this robot running
    local rightMotorTopicName='rightMotorSpeed'..sysTime -- we add a
random component so that we can have several instances of this robot running
```

修改为：

```
    local leftMotorTopicName='leftMotorSpeed'
    local rightMotorTopicName='rightMotorSpeed'.
```

（2）将以下内容：

```
    -- Now we start the client application:
    result=simLaunchExecutable('rosBubbleRob2',leftMotorTopicName..""..
rightMotorTopicName.." "..sensorTopicName.." "..simulationTimeTopicName,0)
```

注释掉，并使用 Lua 注释符”--”注释如下：

```
-- Now we start the client application:
--result=simLaunchExecutable('rosBubbleRob2',leftMotorTopicName.."".. rightMotorTopicName.." "..sensorTopicName.." "..simulationTimeTopicName,0)
```

下面关闭 rosInterfaceControlledBubbleRob 的 Lua 脚本文件，单击“运行”按钮，在可视化区域内可以看到红色小车静止不动。重新打开两个终端，分别运行：

```
rostopic pub /rightMotorSpeed std_msgs/Flt32 "data: 2.0"
rostopic pub /rightMotorSpeed std_msgs/Flt32 "data: 2.0"
```

可以看到，小车接收到 ROS 发送来的速度 Topic 之后开始做直线运动。当然也可以通过给差分轮的左右轮设置不同速度实现转向，如：

```
rostopic pub /rightMotorSpeed std_msgs/Flt32 "data: 2.0"
rostopic pub /rightMotorSpeed std_msgs/Flt32 "data: 0.0"
```

新打开终端，输入 rostopic list 查看所有发布的 Topic：

```
/leftMotorSpeed
/rightMotorSpeed
/rosout
/rosout_agg
/sensorTrigger10864719
/simTime10864719
/tf
/vrep_ros_interface/addStatusbarMessage
/vrep_ros_interface/objectCount
```

新打开终端，输入 rosrun rqt_graph rqt_graph 查看节点关系图，如图 7-11 所示。

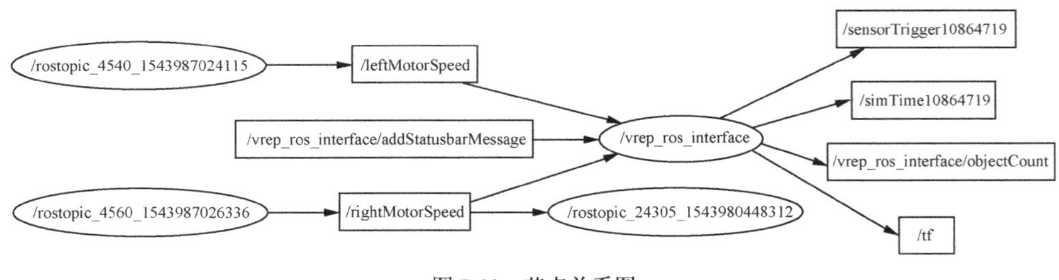

图 7-11 节点关系图

通过节点图可以看出，/vrep_ros_interface 节点接收左右轮的速度 Topic:/leftMotorSpeed,/rightMotorSpeed，并对外发布 /tf 话题。至此，我们已经通过两个自带的场景文件实现了 V-rep 与 ROS 间双向信息交流。为了提升读者的开发能力，这里将 V-rep 中与 ROS 通信的主要 API 函数及功能总结如表 7-2 所示。

表 7-2 主要 API 函数及功能表

API	功能及用法
simExtRosInterface_advertise	Advertise a topic and create a topic publisher simExtRosInterface_advertise(string topicName, string topicType, int queueSize=1, bool latch=false)（订阅主题并创建主题发布者）
simExtRosInterface_publish	Publish a message on the topic associated with this publisher simExtRosInterface_publish(int publisherHandle, table message)（通过发布者在主题上发布消息）
simExtRosInterface_advertiseService	Advertise a service and create a service server. simExtRosInterface_advertiseService(string serviceName, string serviceType, string serviceCallback)（订阅服务并创建服务的服务器）
simExtRosInterface_call	Call the service associated with this service client simExtRosInterface_call(int serviceClientHandle, table request)（在客户端请求服务）
simExtRosInterface_deleteParam	Delete a parameter in the ROS Parameter Server simExtRosInterface_deleteParam(string name)（删除参数）
simExtRosInterface_deleteParamDouble/Bool/int	Retrieve a Double/Bool/Int/String parameter from the simExtRosInterface_deleteParamDouble/Bool/int(string name, double defaultValue=0.0/bool defaultValue=false/int defaultValue=0)（删除布尔型、整型、字符串型参数）
simExtRosInterface_publisherTreatUInt8ArrayAsString	After calling this function, this publisher will treat uint8 arrays as string.Using strings should be in general much faster that using int arrays in Lua. publisherTreatUInt8ArrayAsString(int publisherHandle)（发布者将 uint8 型的数组作为字符串处理）
simExtRosInterface_sendTransform	Publish a TF transformation between frames simExtRosInterface_sendTransform(table transform) Subscribe to a topic（发布坐标系之间的 TF 坐标转换）
simExtRosInterface_subscribe	int subscriberHandle=simExtRosInterface_subscribe(string topicName, string topicType, string topicCallback, int queueSize=1)（订阅主题上的消息）
simExtRosInterface_subscriberTreatUInt8ArrayAsString	After calling this function, this subscriber will treat uint8 arrays as string. Using strings should be in general much faster that using int arrays in Lua.(int subscriberHandle)（订阅者将 uint8 型的数组作为字符串处理）

上述API函数主要实现Topic的订阅、发布、参数获取与删除、发布TF变换等功能。如果读者通过前几章的学习已经具备了一定的ROS开发能力，相信会很快掌握这些函数接口的应用。

> **注意**
>
> 以上函数接口针对旧版本的V-rep，最新的V-rep3.5版本已经将simExtRosInterface_替换成simROS，即simExtRosInterface_subscribe更改为simROS.subscribe。具体内容可查看V-rep官网帮助文档。目前，作者在Vrep3.4版本上使用simExtRosInterface_开头的PAI函数均能正常运行，暂不受影响。

通过以上内容介绍，相信读者已经对V-rep与ROS之间通信有了一定的了解，但是使用官方自带的场景总不如自己搭建一个V-rep场景并进行ROS控制显得有成就感。虽然V-rep自带丰富的机器人和移动平台，但为了真正掌握V-rep与ROS的联合开发，我们有必要亲手搭建机器人及仿真环境。接下来，我们搭建一个属于自己的机器人仿真环境，并进行ROS开发。

7.4 V-rep环境搭建与ROS控制开发

7.4.1 V-rep环境搭建

打开V-rep软件“File → new scene”，新建场景文件，并另存为car_ros_vrep.ttt至合适路径（读者根据自己习惯存放，此处无特别要求）。接下来，制作属于自己的移动小车，单击“Add → Primitive shape → Cuboid”，设置x、y、z三个方向的尺寸分别为0.6、0.4、0.06，如图7-12所示。

单击“OK”，并在Scene hierachy一栏下将“Cubiod”修改为“Car_base”（通过双击“Cubiod”修改）。可以看到，修改后的小车主体如图7-13所示。

单击选中“Car_base”，单击“Object/item shift → position”，修改小车主体的坐标x为0.7、y为-1.7、z为0.15，如图7-14所示。

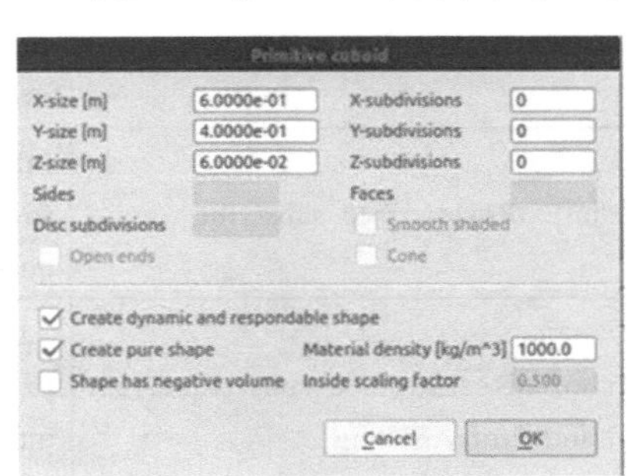

图7-12 设置x、y、z属性

添加小车车轮，单击“Add → Primitive shape → Cylinder”，分别设置 x、y、z 尺寸为 0.3、0.3、0.06，如图 7-15 所示。

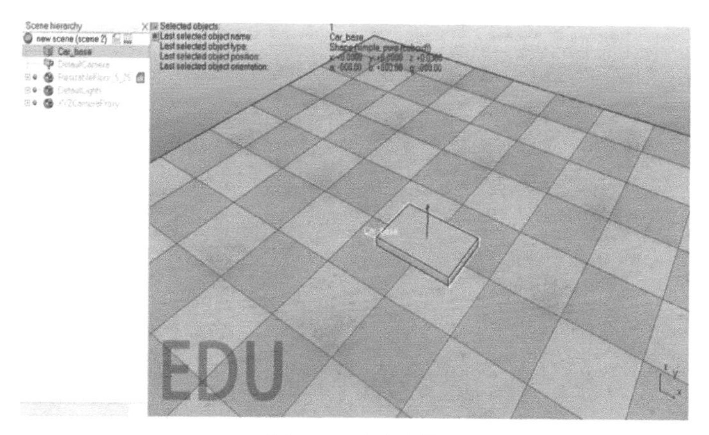

图 7-13　小车主体

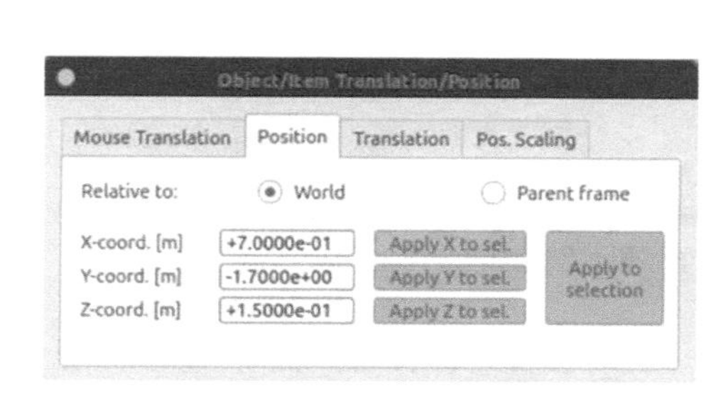

图 7-14　小车主体的坐标

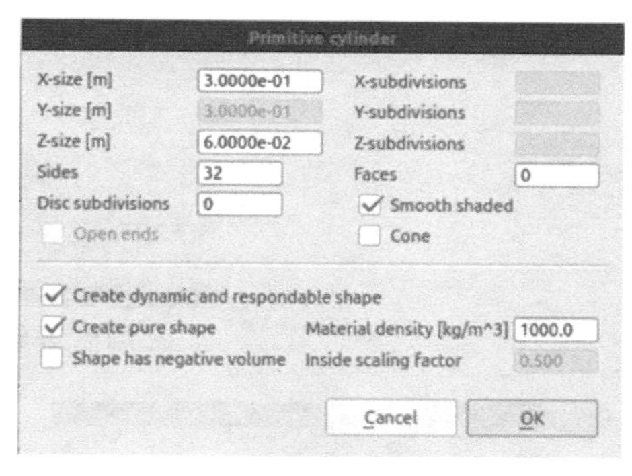

图 7-15　小车车轮尺寸

单击“OK”，同理，重命名为Left_wheel，然后选中左轮并依次单击“Object/item rotate → Orientation”，设置选转角度：Alpha 为 -90、Beta 为 0、Gamma 为 0，如图 7-16 所示。

选中“Left_wheel”，依次单击“Object/item shift → position”，修改左轮主体的坐标 x 为 0.65、y 为 -1.94、z 为 0.15，如图 7-17 所示。

此时，左轮和车体的相对位置如图 7-18 所示。

接下来，为左轮添加动力装置，单击“Add → Joint Revolute”，双击重命名 Left_motor。双击 Left_motor 前的图标设置直径和长度尺寸参数（Length:0.08，Diameter:0.02），如图 7-19 所示。

单击“Show dynamic properties dialog”，勾选并设置 Motor enable 和 Lock motor when target velocity is zero，如图 7-20 所示。

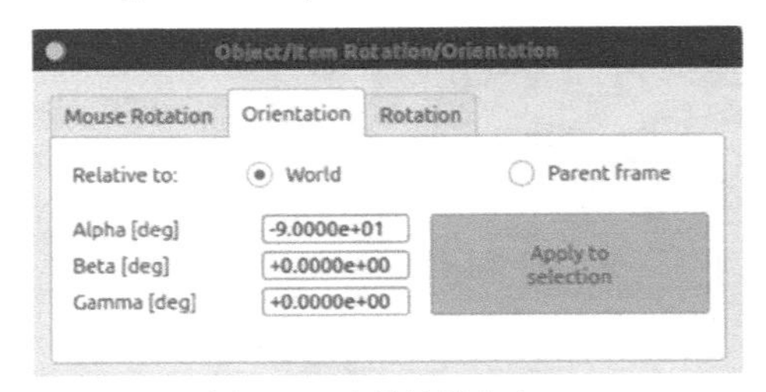

图 7-16　车轮旋转角度

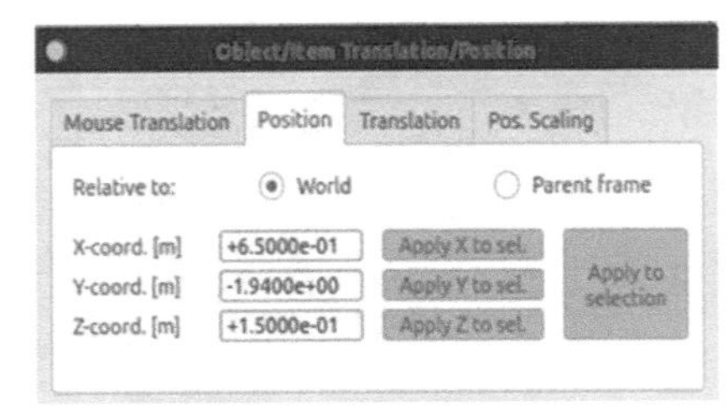

图 7-17　左轮主体坐标

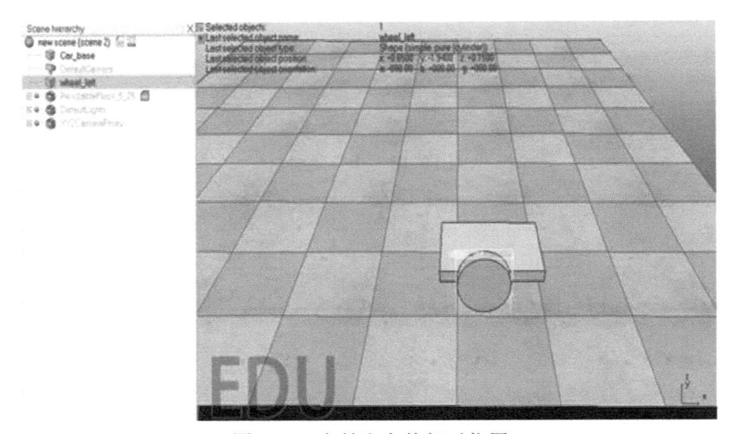

图 7-18　左轮和车体相对位置

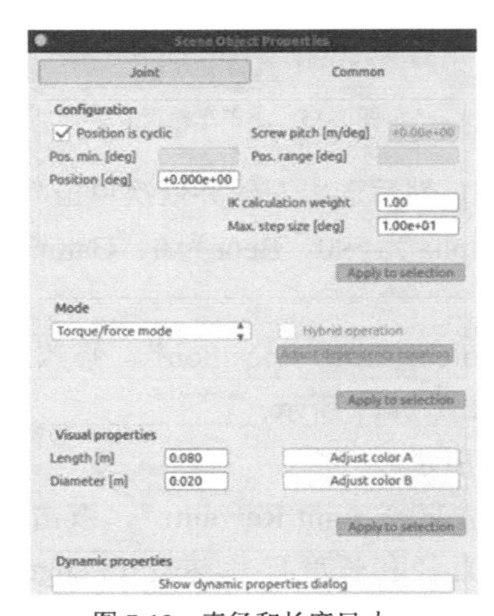

图 7-19　直径和长度尺寸

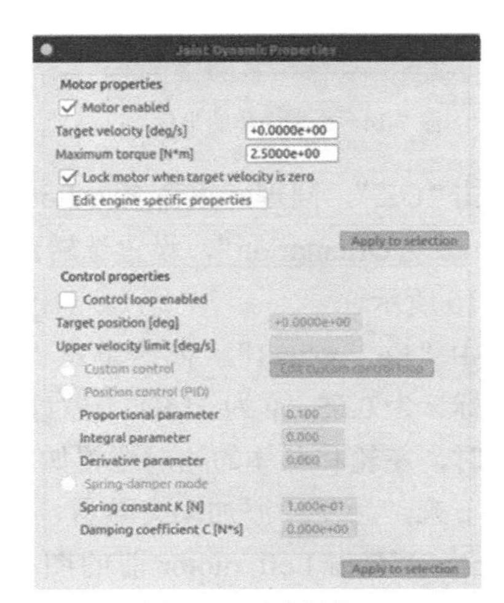

图 7-20　动态属性

选中“Left_motor”，并依次单击“Object/item rotate → Orientation”，设置选转角度 Alpha 为 -90、Beta 为 0、Gamma 为 0，如图 7-21 所示。

选中“Left_motor”，依次单击“Object/item shift → position”，修改左轮主体的坐标 x 为 0.65、y 为 -1.94、z 为 0.15，如图 7-22 所示。

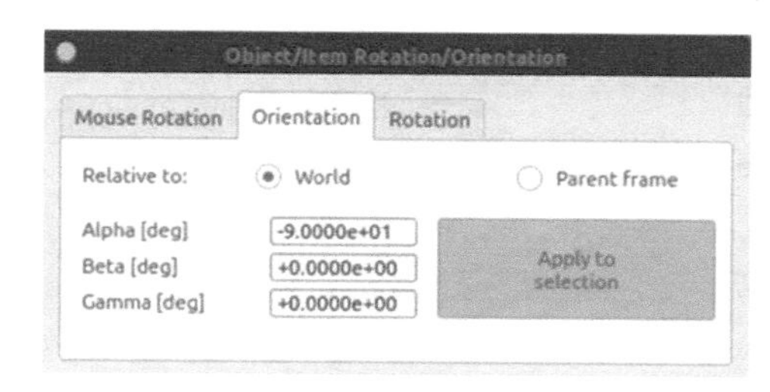

图 7-21 旋转角度

图 7-22 左轮主体的坐标

设置完成后的相对位置如图 7-23 所示。

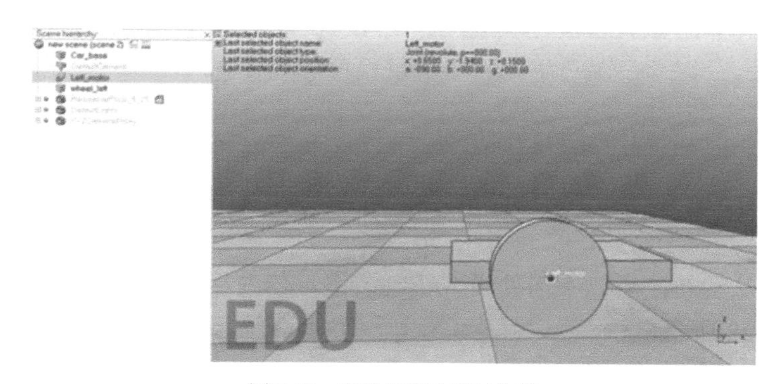

图 7-23 左轮与车体相对位置

同理设置 Right_wheel 和 Right_motor，设置完成后如图 7-24 所示。

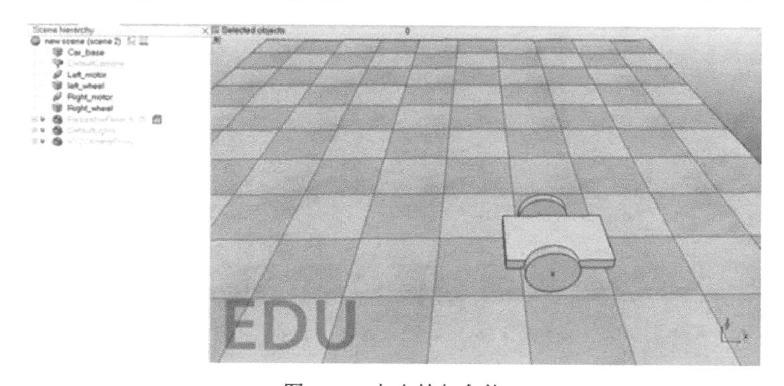

图 7-24 左右轮与车体

下面开始设置小车前轮，依次单击“Add → Primitive shape → Cylinder”，设置 x、y、z 尺寸分别为0.12、0.12、0.04，如图7-25所示。

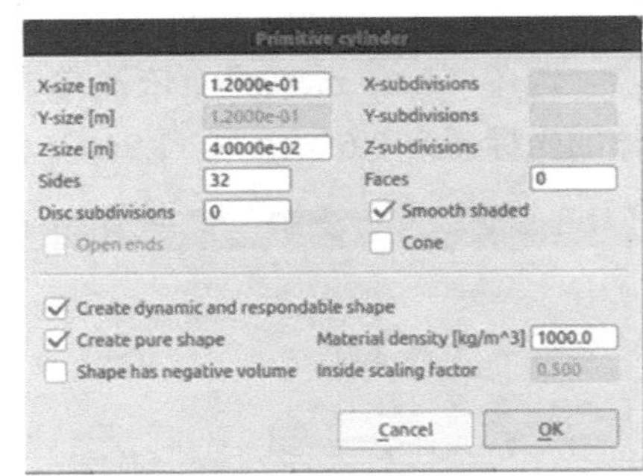

图7-25 小车前轮属性

单击“OK”，同理，重命名为“Front_wheel”，然后选中左轮，并依次单击“Object/item rotate → Orientation”，设置选转角度Alpha为−90、Beta为0、Gamma为0，如图7-26所示。

选中“Front_wheel”，依次单击“Object/item shift → position”，修改左轮主体的坐标 x 为0.9、y 为−1.7、z 为0.06，如图7-27所示。

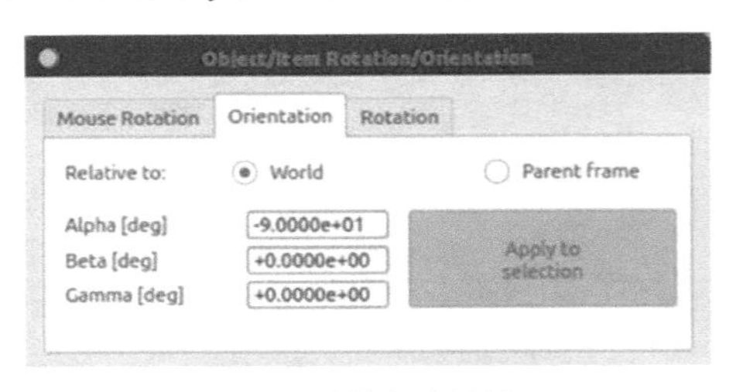

图7-26 旋转角度设置

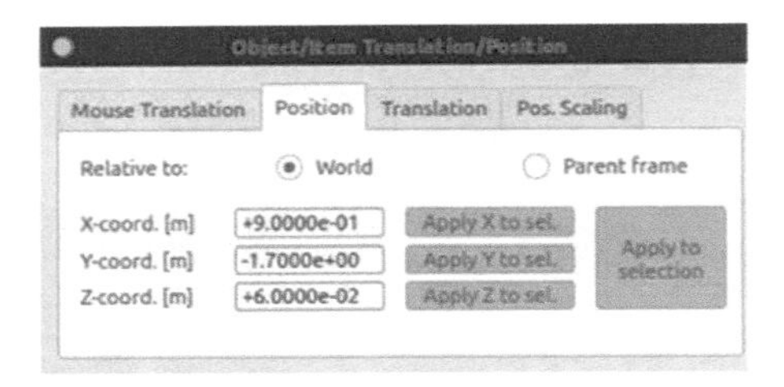

图7-27 位置坐标

由于是从动轮，所以不需要设置驱动装置。修改后的车体及3个车轮的相对位置如图7-28所示。

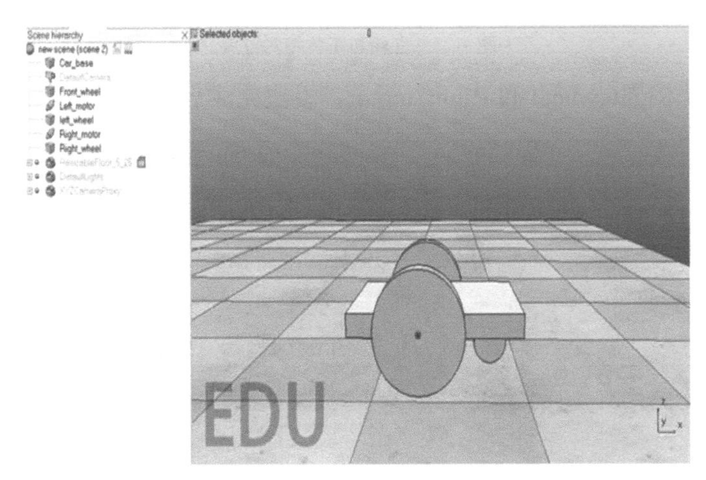

图7-28 车体车轮相对位置

为前轮添加力传感器，单击“Add → Force sensor”，并重命名为“Front_baselink”。选中“Front_baselink”，依次单击“Object/item shift → position”，修改力传感器的坐标 x 为 0.9、y 为 -1.7、z 为 0.13，如图 7-29 所示。

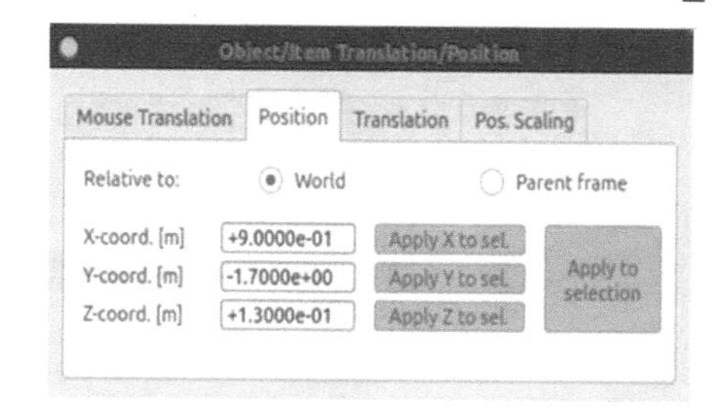

图 7-29　力传感器坐标

此时，前轮加上力传感器的小车如图 7-30 所示（图中前轮上方绿色部分即为力传感器）。

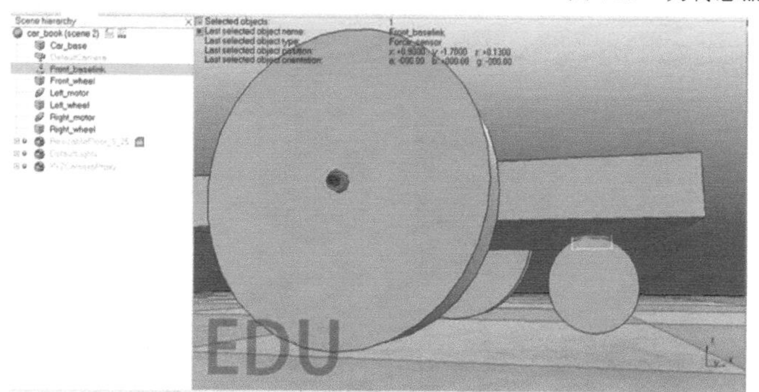

图 7-30　力传感器与车体

通过拖曳使 Scene hierachy 一栏下的 Front_whell 到 Front_baselink 下方，Left_wheel 到 Left_motor 下方，Right_wheel 到 Right_motor 下方。然后分别将 Front_baselink、Left_motor、Right_motor 拖曳到 Car_base 下方，最终形成如图 7-31 所示的树状组织结构。

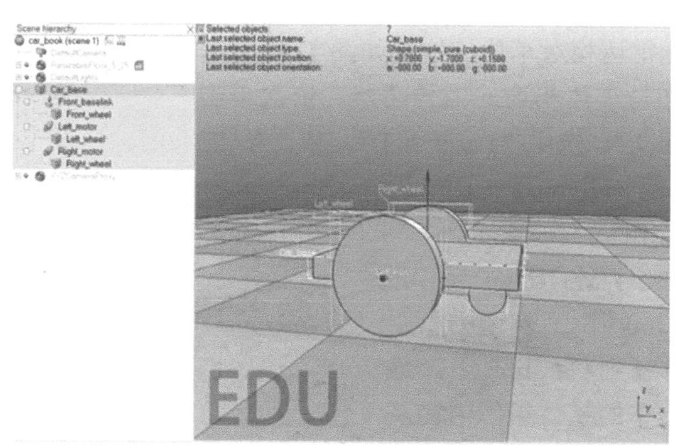

图 7-31　车体组织结构

为小车添加建图导航所用的力传感器，依次单击“Model browser → sensorke”，下拉可以看到许多传感器，如图7-32所示。

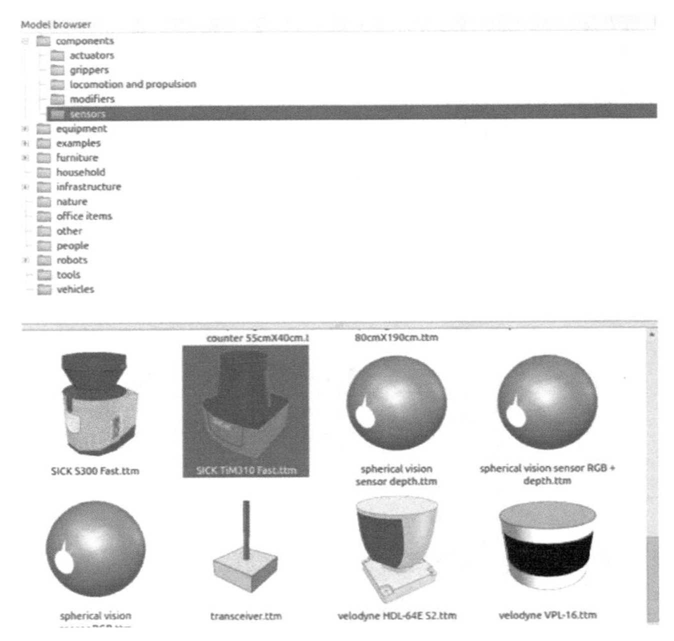

图7-32 V-rep中传感器

这里我们选择SICK_TiM310_fast激光雷达传感器。单击选中，拖曳到右端小车的可视化区域，并通过修改其位置参数x为0.9、y为−1.7、z为0.22，如图7-33所示。

此时，激光雷达相对车体的位置如图7-34所示。

为雷达添加与车体相连的力传感器，依次单击“Add → Force sensor”，并重命名为“Laser_link”。选中“Laser_link”，依次单击“Object/item shift → position”，修改力传感器的坐标x为0.9、y为−1.7、z为0.18，如图7-35所示。

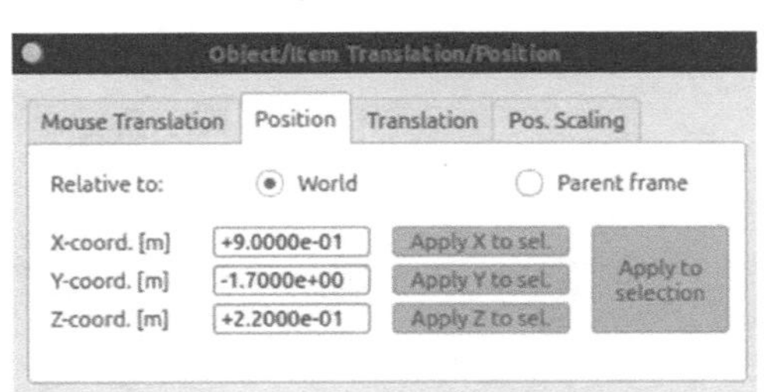

图7-33 激光雷达坐标

选中“SICK_TiM310_fast”，并拖曳到Laser_link下，然后选中“Laser_link”拖曳到Car_link下。最后，我们得到小车的整体组织结构及形状，如图7-36所示。

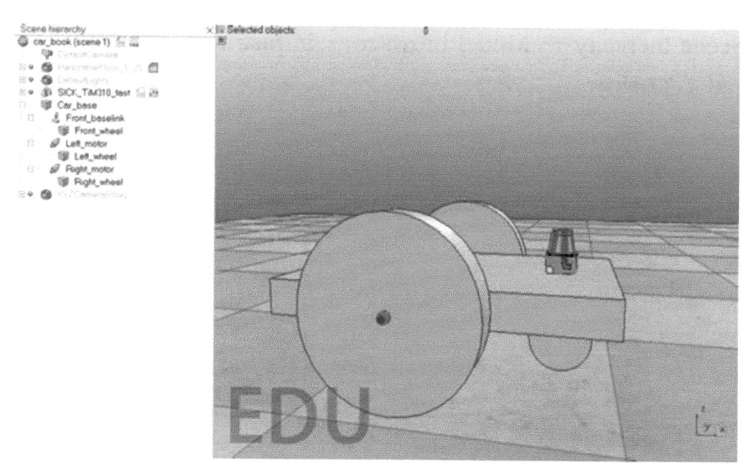

图 7-34 激光与车体的相对位置

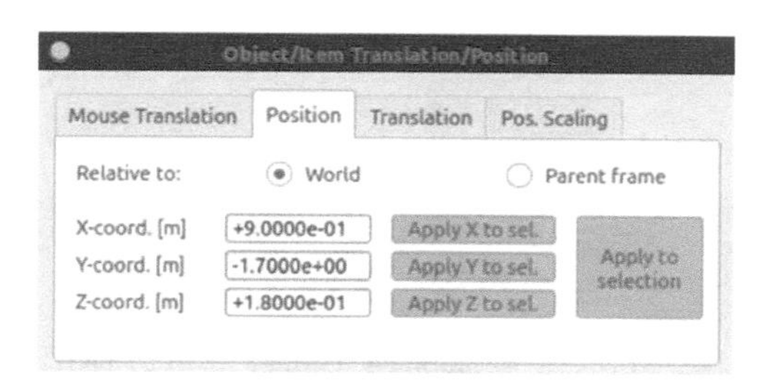

图 7-35 传感器坐标

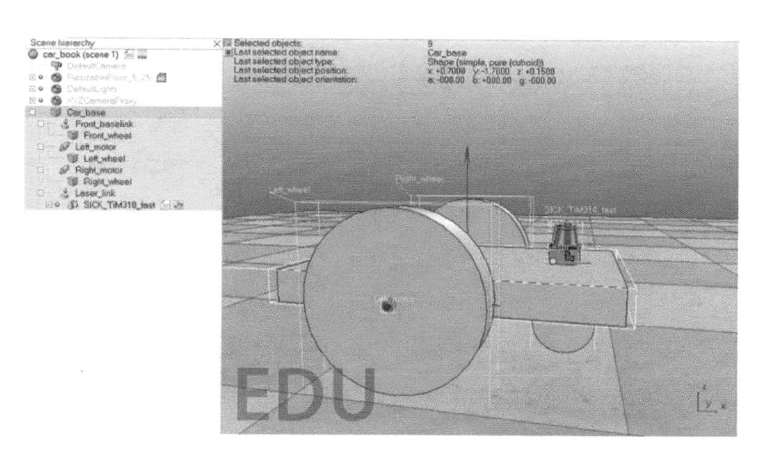

图 7-36 车体组织结构及形状

至此，我们已经完成了小车的模型构建。接下来，搭建小车的运行环境，依次单击“Scene hierachy → Model browser → infrastructure → walls → 80 cm high walls”，如图 7-37 所示。

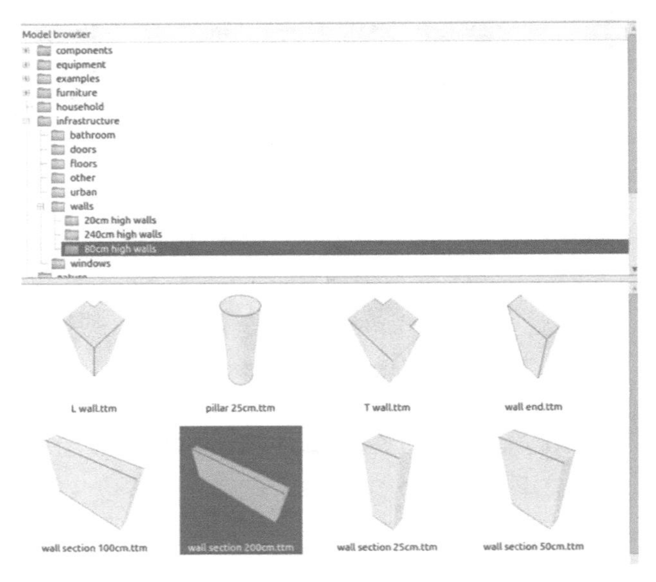

图 7-37 墙体模型

通过多个墙体拼接，并改变墙体选择（读者也可以使用默认颜色，这里没有特殊要求），最终实现小车的环境，如图 7-38 所示。

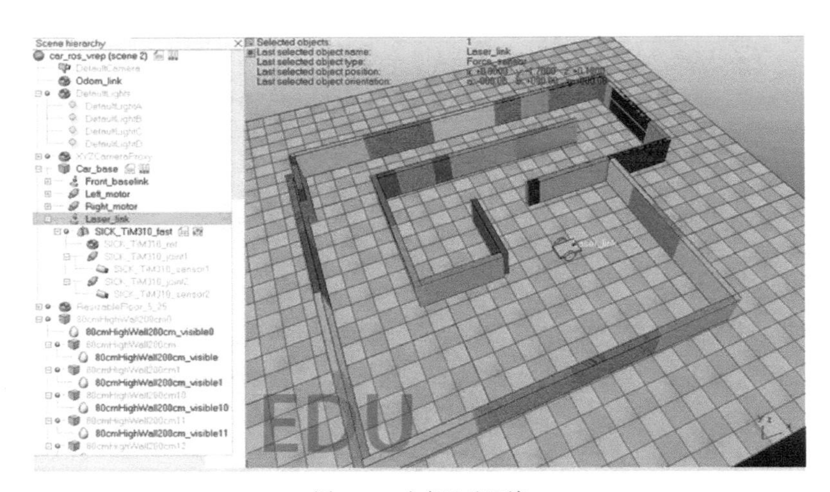

图 7-38 小车运动环境

至此，我们已经完全完成小车及运行环境的搭建。通过单击“仿真开始”按钮，可以看到小车上雷达发出的激光线，如图 7-39 所示。

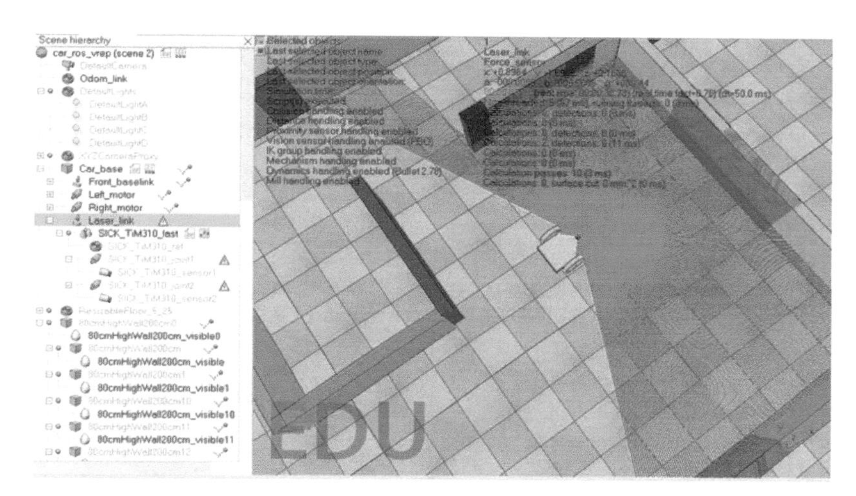

图 7-39　雷达激光线

7.4.2 激光雷达ROS参数配置

以上仿真的激光雷达数据还需封装成 ROS 消息发布出去才能被使用，打开 SICK_TiM310_fast 后面的脚本文件。找到 if (sim_call_type==sim_childscriptcall_initialization) then，在该段找到如下代码：

```
    if (simGetInt32Parameter(sim_intparam_program_version)<30004) then
            simDisplayDialog("ERROR","This version of the SICK sensor is only supported
from V-REP V3.0.4 and upwards.&&nMake sure to update your V-REP.",sim_dlgstyle_ok,
false,nil,{0.8,0,0,0,0,0},{0.5,0,0,1,1,1})
        End
```

在其下方创建 ROS 发布句柄，代码如下：

```
    -- Enable an LaserScan publisher:
        pub = simExtRosInterface_advertise('/scan', 'sensor_msgs/LaserScan')
```

然后，找到文件末端的如下代码：

```
    if notFirstHere then
            -- We skip the very first reading
    ..................
```

```
………………….
notFirstHere=true
```

在此处下方添加封装并发布 ROS 消息的代码：

```
-- Public  the LaserScan message(发布激光雷达消息)
local frame_stamp1 = simGetSystemTime()
scan={}
scan['header']={seq=0,stamp=frame_stamp1, frame_id="base_link"}
scan['angle_min']=angle_min
scan['angle_max']=angle_max
scan['angle_increment']=angle_increment
scan['time_increment']=time_increment
scan['scan_time']=frame_stamp1 -- Return the current ROS time i.e. the time returned
by ros::Time::now()
scan['range_min']=range_min
scan['range_max']=range_max
scan['ranges'] = distanceData
scan['intensities']={}
simExtRosInterface_publish(pub, scan)
```

通过前几章的学习，相信读者已经掌握了 ROS 消息格式，以上代码就是严格按照 ROS 消息格式封装的，区别在于，这里的脚本文件采用的是 Lua 语法。保存文件，进行消息发送测试。单击“运行”按钮，打开终端：

```
rostopic list
```

可以看到已经发布的所有 Topic 名称（其中可以看到激光雷达发布的 /scan），如图 7-40 所示。

若要查看 /scan 的发布者及接收者，可以使用如下命令：

```
rostopic info /scan
```

显示内容如图 7-41 所示。

```
/clock
/cmd_vel
/odom
/rosout
/rosout_agg
/scan
/simTime1728781
/tf
/vrep_ros_interface/addStatusbarMessage
/vrep_ros_interface/objectCount
```

图 7-40　Topic 名称信息

```
Type: sensor_msgs/LaserScan

Publishers:
 * /vrep_ros_interface (http://192.168.0.103:42967/)

Subscribers: None
```

图 7-41　/scan 的发布者及接收者

图 7-41 显示了消息的发布者、接收者及所发布的具体内容。可以看到，/scan 是由 V-rep 端发布的，此时并没有接收 /scan 的节点，Topic 的类型是 sensor_msgs/LaserScan。

打开终端，输入 rosrun rviz rviz 查看可视化的激光雷达信息。在 Rviz 中选择 Fixed Frame:base_link。在 LaserScan 下方选择 Topic: /scan,Style: Points,Color Transformer: AxisColor。可以看到可视化的激光雷达消息，如图 7-42 所示。

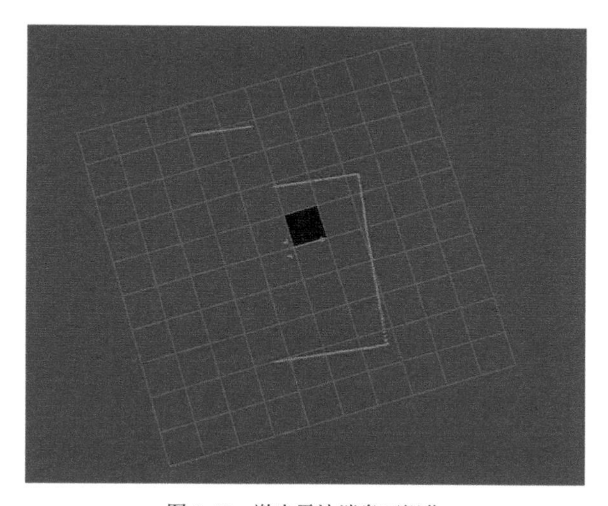

图 7-42　激光雷达消息可视化

至此，我们已经完成了激光雷达在 V-rep 端的发送，以及 ROS 端的接收与可视化。接下来，我们来完成机器人 SLAM 建图及导航实验。

7.5 V-rep与ROS联合仿真实验

7.5.1 gmapping建图测试

此前，我们已经做好了所有的准备工作，包括小车本体建模、运行环境搭建、激光雷达配置等。现在，我们使用 ROS 自带的建图功能包进行 gmapping 建图测试。建图功能包的 Launch 配置文件如下：

```
<?xml version="1.0"?>
<launch>
    <!-- Use the clock time coming from V-REP, otherwise some generated topics
will be outdated as V-REP time is different from the normal timestamp -->
    <param name="use_sim_time" value="true" />

    <node pkg="gmapping" type="slam_gmapping" name="slam_gmapping" args=
"scan:=scan" output="screen"/>
        <node pkg="vrep_odomtf_pub" type="vrep_odomtf_pub" name="odom2base_tf"
output="screen"/>
        <node pkg="tf" type="static_transform_publisher" name="laser2base_tf"
args="0.2 0 0.07 0 0 0 1 /base_link /laser_link 100" output="screen"/>
</launch>
```

注意

为了保证V-rep仿真时间与ROS时间一致，需要在Launch文件中设置use_sim_time为“true”。

接下来打开终端，在Launch文件夹目录下运行：

```
roslaunch vrep_map_building.launch
```

在新终端打开rviz：

```
rosrun rviz rviz
```

可以在rviz中看到产生的地图，如图7-43所示。

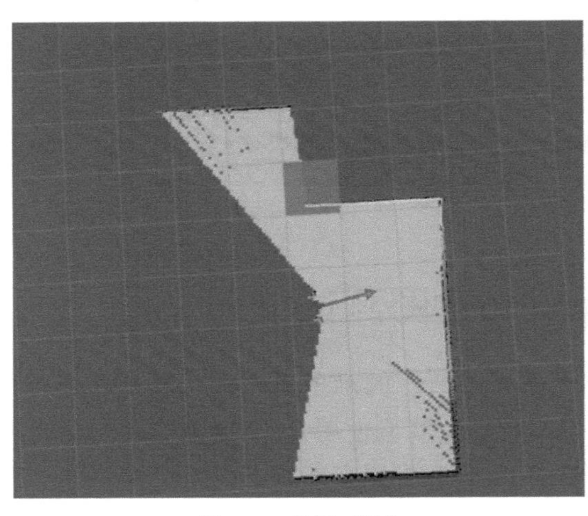

图7-43 地图可视化

新打开终端，通过键盘控制小车运动，输入命令：

```
rosrun  teleop_twist_key  teleop_twist_keyboard.py
```

可以看到如下信息：

```
Reading from the keyboard  and Publishing to Twist!
---------------------------
Moving around:
   u    i    o
   j    k    l
   m    ,    .

For Holonomic mode (strafing), hold down the shift key:
---------------------------
   U    I    O
   J    K    L
   M    <    >

t : up (+z)
b : down (-z)

anything else : stop

q/z : increase/decrease max speeds by 10%
w/x : increase/decrease only linear speed by 10%
e/c : increase/decrease only angular speed by 10%

CTRL-C to quit

currently:    speed 0.5    turn 1.0
```

根据上述提示信息，可通过 <i> 和 <,> 键控制小车的前后运动，<j> 和 <l> 控制左转、右转。若要调整速度，可使用 <q> 和 <z> 键。不断控制小车运动，直到将运行环境行走一遍，并得到满意的地图（若有些区域地图不完整，可重复扫描）。新打开终端，通过下面的命令保存地图：

```
rosrun map_server map_saver -f /path_to_your/xxx
```

其中，xxx 为地图文件名，前方是读者自定义的保存路径。构建完成的完整地图如图 7-44 所示。

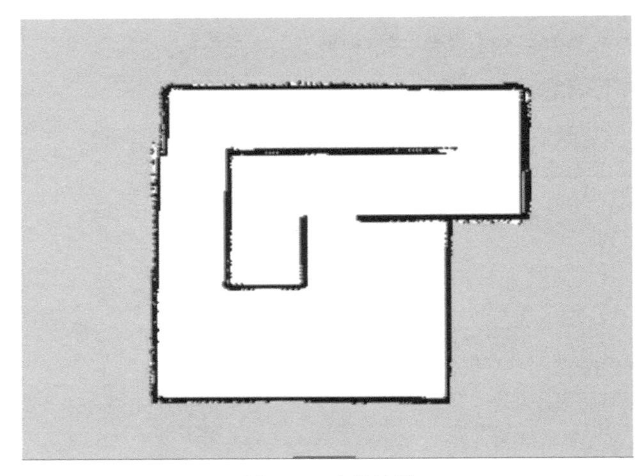

图 7-44 完整地图

7.5.2 导航测试

我们已经构建了机器人导航所用的全局地图。根据前几章的知识，相信读者已经能够完成 ROS 的 navigation 文件参数配置。参数配置完成之后，根据导航启动文件即可进行导航测试，测试步骤如下。

（1）启动 ros master：

```
~$  roscore
```

（2）启动 V-rep 软件，加载我们构建的机器人仿真环境，单击“仿真开始”按钮进行仿真。

（3）启动导航测试 Launch 文件：

```
~$  cd ~/path_to_your_navigation_launch
~$  roslaunch navigation.launch
```

（4）打开 rviz：

```
~$  rosrun rviz rviz
```

在 rviz 中根据 Topic 选择 Map、Path、Odometry、LaserScan、Polygen 等 Topic 并重命名，可以看到如图 7-45 所示的代价地图。

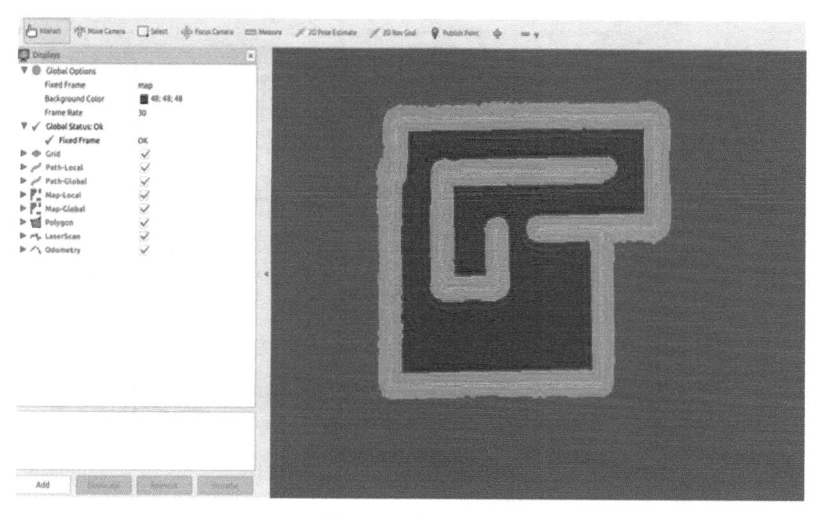

图 7-45 代价地图

单击右上角的“2DPose Estimate”，将机器人的初始位置摆正，如图 7-46 所示（其中，红色矩形代表小车形状）。

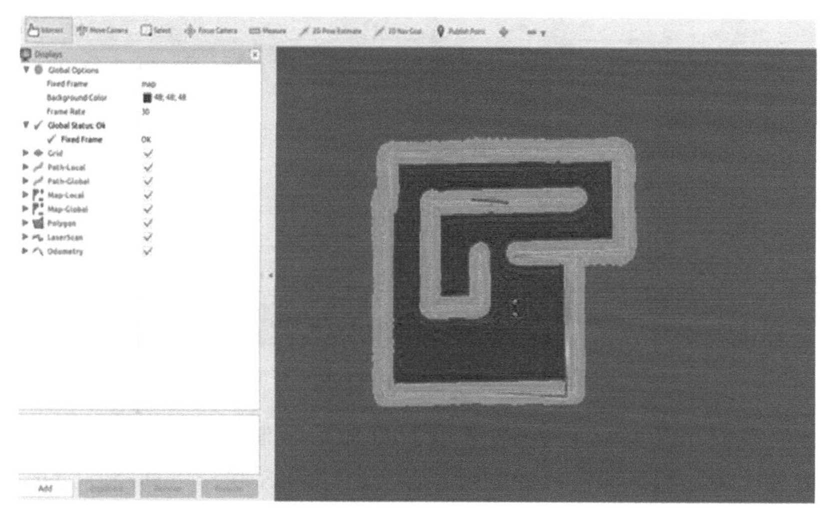

图 7-46 小车位置初始化

单击右上角的“2D Nav Goal”，设置目标点的位置及方向，可以看到自主规划的全局路径及局部路径，如图 7-47 所示。

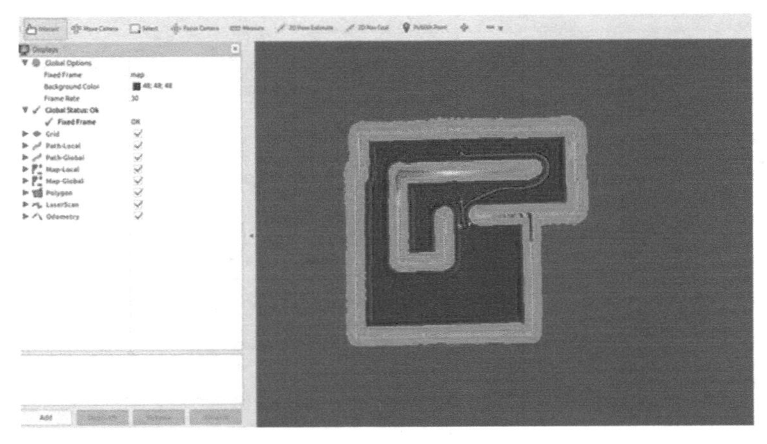

图 7-47　自主导航路径规划

小车在 V-rep 中的运动情况如图 7-48 所示。

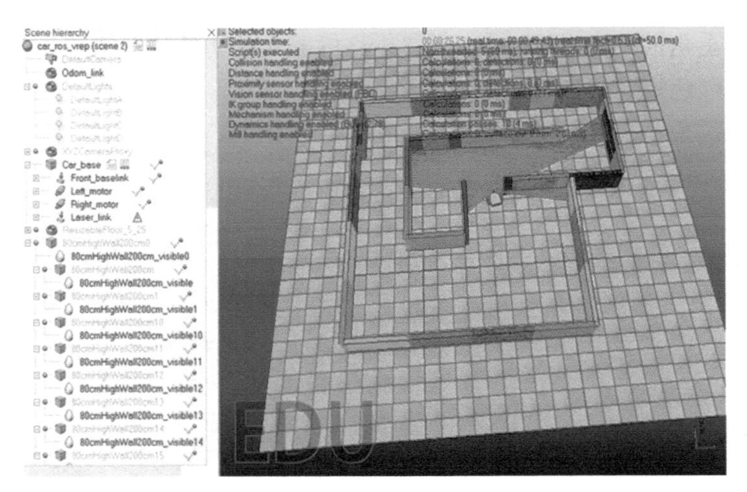

图 7-48　小车在 V-rep 中的运动情况

至此，我们已经完成了 V-rep 环境下的 ROS 控制开发。通过搭建自己的机器人及仿真环境，相信读者已经掌握了 V-rep 与 ROS 联合仿真控制的基本知识。本书为没有机器人实物或实验环境的研究者提供了便捷的开发工具，以后我们不必亲手搭建小车或机器，直接使用 V-rep 自带的机器人即可。当然读者也可以从网上下载相关机器人仿真环境或机器人模型。我们可以将更多的精力投入算法研究中，而不是用在模型构建或传感器搭建上。